ENDANGERED SPECIES

Asia II

Grolier Educational Corporation
SHERMAN TURNPIKE, DANBURY, CONNECTICUT 06816

Grolier Educational Corporation

Remmel Nunn, Vice President, New Product Development
Beverly A. Balaz, Vice President, Marketing
Robert R. Hall, Vice President, Sales
Sara Jane Kennerley, Managing Editor Canadian Division

The Grolier World Encyclopedia of Endangered Species

Remmel Nunn, Project Director
Claudia Durrell and Beverly A. Balaz, Cover Designers
Michael Aslett, Production Coordinator

Published 1993 by Grolier Educational Corporation

Edited, updated, typeset, and redesigned by Grolier Educational Corporation, 1993. Cover design
   © Grolier Educational Corporation, 1993. Photography credits appear separately.
   © Kodansha Ltd., MCMLXXXVIII (1988).

Library of Congress number: 92-054833
Sekai no tennen kinenbutsu. English.
   The Grolier world encyclopedia of endangered species.
      p. cm.
   Contents: 4. Asia II.
ISBN for full set: 0-7172-7192-7 (reinforced-library binding)
1. Endangered species—Encyclopedias. 2. Wildlife conservation—Encyclopedias.
I. Grolier Educational Corporation. II. Title.
QL82.S4513 1992   92-54833
591.52'9'03—dc20   CIP
                                                                      AC

Printed and bound in the United States of America.

Editor's Preface

The Grolier World Encyclopedia of Endangered Species is based on the *Red List* (or *Red Data Book*), a comprehensive and authoritative list of threatened species compiled by the International Union for the Conservation of Nature and Natural Resources (IUCN). Every species listed in the *Red List* as "Rare," "Vulnerable," "Endangered," or "Extinct" can be found in this encyclopedia.

Because the environmental, political, and sociological forces that threaten these species are generally regional in nature, this encyclopedia is organized by geographical region. Within each region, entries are subdivided into classes, orders, and families, following established scientific order. Entries are accompanied by maps showing current distribution and charts indicating the level of threat that the species faces. The text focuses on each species' natural habitat, physiological characteristics, and customary feeding, mating, and migratory patterns. Also treated are the factors that threaten the species, and what actions are being taken to save it.

It took 4 years to commission and collect the more than 700 color photographs in this encyclopedia, and they are central to its purpose. Special emphasis was placed on using photographs of surviving species in their natural habitat, rather than in captivity. Seeing them as they exist in the wild is in itself educational, because it drives home a point that words alone cannot convey: These creatures are actually out there, facing a bleak present and a fragile future.

Categories of Endangerment

The categories of endangerment used in this work are those set by the IUCN, which are recognized internationally. When the classification is uncertain because of lack of information, more than one category has been specified. The abbreviations used in this series are explained below.

(Ex) *Extinct*
Species not identified in the wild in the past 50 years.

(R) *Rare*
Species with small populations, which, although not currently considered "Endangered" or "Vulnerable," are at risk. Generally those animals are found in limited territories or habitats or have a low population density over wide areas.

(E) *Endangered*
Species in danger of extinction and whose survival seems improbable if the factors that have put them at risk continue to affect them. Species whose numbers have decreased to a critical level or whose habitats have been drastically reduced are included in this category. Animals that may possibly be extinct but that have been identified in the wild in the last 50 years are also included.

(T) *Threatened*
Species considered "Endangered," "Vulnerable," "Rare," "Indeterminate," or "Insufficiently Known" when not enough is known about them to place them in one of those categories with certainty. In

this volume this term is also used when the subspecies of a species belongs to different categories.

(V) *Vulnerable*
Species that will be considered "Endangered" in the future if the factors that currently threaten them continue to affect them. Also included are those species whose population, or a large part of whose population, has been reduced because of overexploitation, extensive destruction of habitat, or any other environmental disturbance; animals whose populations have been greatly reduced and whose survival is not guaranteed; and species whose populations, although large, are threatened by serious adverse conditions throughout their habitat.

(I) *Indeterminate*
Species that belong in the categories "Endangered," "Vulnerable," or "Rare," but about which not enough is known to label them exactly.

(K) *Insufficiently Known*
Species believed to belong to one of the categories above but about which reliable information for making that judgment is lacking.

Table of Contents

PYGMY HOG

(Sus salvanius)

Ex	R	E	T	V	I	K

In March 1971 the members of a scientific expedition in India had the extraordinary opportunity of photographing, in the Manas Sanctuary in Assam, at the foot of the Himalayas, a pygmy hog *(Sus salvanius)*. For many years this animal had been believed extinct.

This very small animal is only 25 to 30 cm (10 to 12 in) in height at the withers and is not more than 60 cm (23 in) long. Its fur is dark brown, tending to rusty red, and on the back the bristles form a small mane. The shape of the body is rather squat, whereas the snout is sharp; the ears and tail are short and hairless. The females of the species have only three pairs of mammae, unlike the other species of the genus *Sus*.

The habitat of the pygmy hog is the jungle and the so-called thatchlands (bushy grasslands of plants that are used for the building of house roofs), which are to be found on the southern slopes of the Himalayan range. These grasslands offer shelter to another animal in serious danger of becoming extinct: the hispid hare *(Caprolagus hispidus)*.

The first report of the existence of the *Sus salvanius* goes back to 1847. At present, its area of distribution is very small and covers only the forest strip of northern Assam of the Bhutan border, including the protected area of the Manas Wildlife Sanctuary. It was also probably seen once in Cegalaya, south of the Brahmaputra River, but it does not seem to be very probable that it crossed the river barrier. In the past, however, this animal's distri- bution covered a wider zone of northern India: Sikkim, Nepal, northern Bangladesh, Bhutan, and Bengal.

Not much is known of the habits of this wild boar, due to both its extremely wary character and the now very limited size of its populations, believed to be reduced to not more than 100 specimens. It is known that this animal lives in groups of about 15 to 20 specimens and that the adult males bravely defend the herd from any attacks or intrusions by throwing themselves against the disturbers at a very high speed, often wounding them with their small, sharp tusks.

During the many fires that take place in the underbrush during the dry season, the herds have been seen to run away with extreme rapidity, always remaining united. These mixed groups separate only in January for a short period during the mating season.

At the end of gestation, which goes on for approximately 4 to 5 months, from three to six young are born. They are very small and pretty, with the back longitudinally striped with yellow-brown bands.

The diet of the pygmy hog does not differ much from that of the other animals of the family Suidae. In fact, it is composed of vegetative food, such as roots, berries, and tubers, alternated with insects and other small animals.

This species faces the immediate danger of extinction, mainly because of the destruction of its habitat in consequence of the increasing human settlements (immigration, especially from Nepal, is constantly growing). Especially over the last 20 years, the thatchlands have been subject to tremendous agricultural and commercial pressure, because the dominant grass of these lands is the raw material for the building of house roofs. Consequently, the ground is periodically burnt to improve the harvest. These fires, which are regularly permitted, deprive the wild boar of its food and compel the animal to seek shelter in nearby cultivated areas. These are mainly tea plantations, and these animals are driven away because of the damage they cause to the cultivations.

The pygmy hog is cited in the Washington Convention as well as in the first list of the Act for the Protection of Nature promulgated in India in 1972, which is supposed to put a stop to indiscriminate hunting. Unfortunately, the law is made inef-

The habitat of the pygmy hog is the jungle and bushy grasslands on the southern slopes of the Himalayan range.

fective by a shortage of communication and control in the area concerned.

The various attempts to breed the pygmy hog or have it reproduce in captivity have nearly always been unsuccessful, but more careful and organized attempts might represent a further assurance for the species.

<table>
<tr><td colspan="7" align="center">JAVAN WARTY PIG
(Sus verrucosus)</td></tr>
<tr><td>Ex</td><td>R</td><td>E</td><td>T</td><td>V</td><td>I</td><td>K</td></tr>
</table>

| Ex | R | E | T | V | I | K |

JAVAN WARTY PIG

(Sus verrucosus)

This pig gets its name from the three large warts on each side of its head, one behind the corner of the mandible, one above the eye, and one under it. It weighs at most 150 kg (330 lb), measures about 100 to 160 cm (39 to 62 in) in length, and is 85 cm (33 in) high at the withers. Its coat is sometimes rather thick, mainly brown, with a remarkable variability in color. It can live in various environments, from humid, swampy areas to mountainous slopes, but in general it prefers shaded areas with abundant vegetation.

It is found in Java, Madura (where it is almost extinct), Sulawesi, Flores, the Philippines, and in other small Indonesian islands, where it has given rise to at least 11 subspecies.

This pig prefers to live in small herds or in small family groups. Like other wild boars, it is omnivorous, has an excellent sense of smell, and moves about in the tangle of vegetation with surprising agility. The young are rather robust at birth and are soon capable of following the mother in search of food and are ready to discover the surrounding world. The warty pig is included in the IUCN *Red List* as a species in danger of extinction.

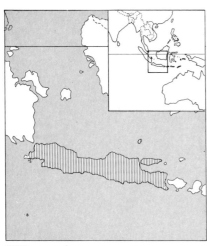

Distribution area of the Javan warty pig

SUMATRAN RHINOCEROS

(Didermoceros sumatrensis)

The Sumatran rhinoceros is both the most primitive representative of the family Rhinocerontidae and the smallest of all the living rhinos. Measuring at most 2.8 to 3 m (9 to 10 ft) in length and 1.4 m (4 1/2 ft) in height at the withers; it rarely weighs more than one ton. It is more similar to the African rhinos than the Asian ones. It has two horns, the front one of which measures 45 cm (18 in) in the males but is reduced to 15 cm (6 in) in the females. The back horn is always much shorter, and females have only the beginning of one. Most of the body is sparsely covered with blackish-gray hairs, which grow thinner almost to the point of disappearing in the course of the years. The tail has long hairs, too, and the ears are bordered by a fringe of bristles.

The Sumatran rhino does not have the typical "carapace" of the Asian rhinos, because the folds of its skin are much less deep, and the cutaneous surface, devoid of tubercles or bonplates, appears much smoother.

Like all rhinos, it prefers swampy areas, rich in water and reeds, but its habitat comprises a great variety of types of vegetation, from the typical humid tropical forest to the drier deciduous forest. The Sumatran rhino, in fact, moves within very wide territories, wandering from the plains to altitudes of up to 2000 m (6500 ft).

Once this species was to be found all over Southeast Asia, from Assam to Vietnam and the Malaccan peninsula as far as the islands of Sumatra and Borneo. Today, its distribution is limited to small areas in Thailand, Burma, Malaysia, Sumatra, and Borneo (at Sabah). In 1983, five rhinos were seen by natives at Sarawak (in Borneo), where they

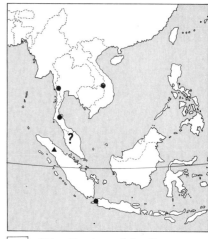

Distribution of the Sumatran rhinoceros:
Documented area of presence
Areas where presence is doubtful

had been considered extinct for 40 years. The animals were in an isolated valley, the exact location of which has not been revealed for safety reasons, and the news of their finding was made known only in 1987.

There are three subspecies of *Didermoceros sumatrensis: D. s. sumatrensis, D. s. harrissonii,* and *D. s. lasiotis.* The Sumatran rhinos have mainly nocturnal habits; they rest in shady places or roll in the mud of bogs during the hottest hours of the day, while at night and in the first hours of the morning they wander about in search of food. The principal element of their diet is not grass but various species of arboreal plants, of which they avidly eat the leaves and tender branches. They also seem to have a very strong predilection for fallen fruit.

They lead a rather solitary life, wandering from one part of their vast territory to another. They deposit their excrement in heaps along the paths or near the swamps or water courses that they usually frequent. The eyesight of these shy animals is rather weak, but hearing and sense of smell are very acute. It is very hard to approach a rhino in the thick of the forest without being noticed.

The Sumatran rhino reaches sexual maturity at between 3 1/2 and 6 years of age. The female

gives birth to only one offspring every 3 years. A definite mating season does not seem to exist, and the solitary life that this animal leads, together with the vastness of its territory, does not favor mating. Today, the repro-

ductive potential of the species is made even weaker by the fact that the populations are extremely reduced or fragmented. In the Malaysian peninsula, for instance, in 1984 there were from 50 to 75 specimens irregularly

The Sumatran rhinoceros has two horns, like the African rhinoceros. In the females, the rear horn is barely developed. The species is predominantly nocturnal.

distributed. The largest group was in the area of Endau Rompin and was composed of no more than 20 animals. The situation of the species is therefore very critical, and it seems that the number of surviving Sumatran rhinos is not more than 100.

The causes for this are to be found in the indiscriminate hunting of past centuries which still continues today despite prohibitions. The horn and other parts of the animals have a high commercial value because of their supposed medicinal properties. The already scanty populations are further diminished by the agricultural exploitation of the territories they frequent and by deforestation, which, in a few years, may well completely destroy the habitat of the Sumatran rhino. New natural reserves and national parks, big enough to protect the animal in its wanderings (like the one proposed at Endau Rompin), need to be instituted. The already existing reserves in Malaysia, Thailand, Sumatra, and Sabah ought to be better protected.

The species *Didermoceros sumatrensis* has been included in the list of the Washington Convention.

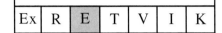

JAVAN RHINOCEROS

(*Rhinoceros sondaicus*)

Ex	R	E	T	V	I	K

The Javan rhino is a very rare animal, probably one of the rarest of the large mammals. It is vary similar to the Indian rhino of the same genus, except for its smaller size and the different conformation of the folds of the hide, which, in the *Rhinoceros sondaicus*, join together behind the shoulders. The size of the horn is smaller, too: only 25 cm (10 in) in the male, whereas in the female it is almost entirely absent. The body is about 3 m (10 ft) long, including the tail, and it varies from 140 to 170 cm

The Javan rhinoceros is a solitary animal that lives where the vegetation is thickest in swampy areas and humid tropical forests.
This species has become very rare; its population is estimated at just over 50 specimens (1986).

(55 to 66 in) in height at the withers. The skin, of a brownish gray color, looks like a carapace formed of large plates, in their turn divided into small polygonal scales. There are hairs only around the ears and on the tip of the tail. Its head is slightly more elongated in comparison with that of the Indian rhino, and the upper lip is elongated to form the typical finger like appendix.

The Javan rhino prefers environments with very thick vegetation, swampy areas, and humid tropical forests. Although it does not seem to be particularly fond of high altitudes, it often ventures into the mountains and along volcanic slopes to avoid human presence. Once it frequented the plains, in jungle clearings, but this habit made it extremely vulnerable to humans.

Originally present in part of India, in Burma, Vietnam, Thailand, Bangladesh, Kampuchea (Cambodia), Laos, Malaysia, and the islands of Java and Sumatra, the *Rhinoceros sondaicus* is present today only in the Udjung Kulon National Park, in the extreme west of Java. This protected area began as a hunting reserve in 1931 and 60 years later was declared a national park. Besides

the peninsula of Udjung Kulon, it includes two small northern islands and the reserve of Gunung Honje further east, but the rhinos live only in the 30,000 hectares (74,000 acres) of the peninsula. This is one of the most interesting parks in the world, both for its flora and for its fauna. Most of it is covered with very thick semideciduous forest, and it has a marine climate with an average yearly rainfall of 3000 mm (117 in).

Not much data is known about the habits of the Javan rhinoceros. It is a rather solitary creature, feeding mainly on shrubs, shoots, and leaves. It becomes independent when quite young and does not seem to have a definite mating season. Its reproductive potential is low, and the female's heat lasts only one day.

The population of the *Rhinoceros sondaicus* was estimated at only 54 specimens in 1986. This animal was already very rare in the 1930s because of ruthless hunting, encouraged by the belief that every part of its body possessed great therapeutic properties. Extracts and powders obtained from the skin, meat, blood, urine, and above all from the horn constituted the basic ingredients of antidotes for epidemics and many other ailments. In Sumatra, it was believed that if a poisoned drink were poured into a cup made of rhinoceros horn, the liquid would foam, and the powder obtained from the horn, mixed with water, was considered a powerful aphrodisiac.

Twenty-five years ago only about 20 Javan rhinos remained because of the destruction of their natural habitat by humans. Since then, various measures have been taken to save this species from extinction. In 1967, the ethologist Schenkel carried out studies on the species and, in that same year, the WWF began to furnish concrete aid to the Indonesian authorities so as to take effective measures against poaching.

Today, there is a problem of overpopulation in the park of Udjung Kulon. The animals, constricted in a space too small for them, weaken genetically, and their resistance to disease is greatly reduced. In 1982, for example,

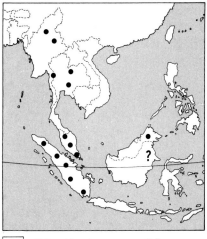

■ • □ *Distribution area of the Javan rhinoceros*
Recently confirmed presence
□ ? □ *Doubtful presence*

five rhinos died within a short time, probably because of an epidemic. To resolve this worrying situation, it has been proposed that a few specimens be moved to a suitable area where they could live and reproduce, thus creating a new colony. Some scholars have advised waiting until the number of rhinos at Udjung Kulon increases from 54 to at least 80 individuals and the technique of reintroduction becomes safer and more reliable, so as not to run risks. The Javan rhinoceros is among the species listed in the Washington Convention.

GREAT INDIAN RHINOCEROS
(Rhinoceros unicornis)

Ex	R	E	T	V	I	K

The Indian rhinoceros, the largest of the three Asian rhinos, measures 4 m (13 ft) in length, including the tail, and varies in height at the withers from 1.6 to 1.9 m (63 to 75 in). The body is massive and weighs about two tons, sometimes more. It has a single horn at the extremity of the snout that is up to 53 cm (21 in) long. The head is rather large, with a greatly developed nuchal crest especially in the male. The upper lip has been transformed into a finger like appendix, very useful for cropping buds and twigs. The lower canine teeth are very strong, whereas the incisors appear smaller and sharper. The very thick skin of the Indian rhinoceros is furrowed with deep wrinkles that divide it into wide areas, similar to armor plates because of the presence of round, horny protuberances. There is no hair, except for a tuft of bristles at the tip of the tail

The very thick skin of the great Indian rhinoceros is furrowed with deep wrinkles that divide it into wide areas similar to armor plates because of the presence of round, horny protuberances.

and a fringe of hair that borders the ears.

It is brownish gray in color, slightly pink near the hide's plicae. The legs end in three toes with wide, flattened hooves. Despite its enormous body structure, this animal is capable of moving with surprising speed and elegance: launched at a gallop on suitable ground, it can reach 35 to 40 kmph (22 to 25 mph).

The favorite habitat of the Indian rhino is swampy alluvial plains crossed by numerous water courses, rich in high grass, reeds, and bogs. These animals often go into drier, cultivated zones, much to the anger and annoyance of farmers.

Five hundred years ago the area of distribution of the species extended over a vast territory comprising the alluvial plains of the Ganges, Indus, and Brahmaputra rivers, bordered on the west by the mountainous zones of the Hindu Kush and on the east by Burma. The northern boundary was constituted by the last Himalayan buttresses. At present, the Indian rhino may be found in very limited and fragmentary areas in the Brahmaputra Valley, in western Bengala, and in the region of Chitawan in Nepal.

Like other rhinos, this species is herbivorous; its diet consists 80 percent of herbaceous plants and, for the remaining 20 percent, of reeds, leaves, shoots, and swamp plants of different kinds, according to seasonal availability. It feeds in the morn-

ing and evening, dedicating the rest of the day to rest.

It is a rather solitary animal and only rarely forms groups of four to six individuals, generally composed of young males. It is not at all aggressive and if disturbed prefers to flee, unless it is a female with its young or a wounded specimen. In these cases it can react violently, charging at the intruder with head lowered. However, contrary to common belief, the rhino does not strike its adversary with its horn, but with the lower, very sharp incisors.

It takes mud baths, which are very useful for defense against parasites and for reestablishing thermal balance. In the hottest hours it is easy to come across a group of a few individuals, com-

pletely immersed (except for the head) in a river or pool.

For social communication, these rhinos use at least 10 different types of sounds and, above all, olfaction; they deposit heaps of excrement around their resting places, feeding grounds, and paths to mark their territory.

In general, they are not territorial animals in a strict sense, but they can become so on particular occasions, such as during the mating period, which lasts from February to April. Gestation lasts about 16 months. The young rhino, hornless, weighs 60 kg (132 lb) at birth.

With its enormous body and menacing appearance, this pachyderm has no natural enemies except the elephant and, of course, man, who hunts it persis-tently for its horn, to which extraordinary aphrodisiac and medicinal properties are attributed in the Asian countries. An illegal organization in Assam controlled the poaching. An intermediary would hire two or three men to kill the rhinos, furnishing them with the necessary weapons and paying them, on delivery of the horn, from $1200 to $2600 a kilogram (2 lb). Afterward, the precious trophy, by way of Calcutta, was sent to Singapore, where, until 1986, most rhino horns ended up, for there were no restrictions on its importations. Here, its price rose to $9000 a kilogram. In this way, from 1981 to 1985, an average of 31 rhinos a year were killed in the national park of Kaziranga.

Because of hunting, at the beginning of this century the species had become almost extinct in India. Today, about 4700 specimens still survive, most of which are in the 430 square km (166 square mi) of the Kaziranga National Park in Assam (1080 head) and in the 900 square km (347 square mi) of the Royal National Park of Chitawan in Nepal (350 head). To stop the slaughter would require twice the number of for-

The great Indian rhinoceros is a primarily solitary animal; it forms groups of four to six individuals, usually young males. It uses at least 10 different vocalizations for group communication.

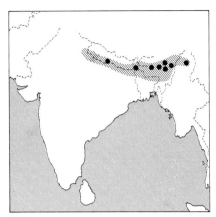

Distribution area of the great Indian rhinoceros

▦ Former distribution area
● Current distribution area

The wild Bactrian camel is found in arid zones of central Asia. This is a rare photograph.

est guards in Assam (increasing the control in the national parks, above all in the border areas). It would be necessary to enforce the laws on poaching more severely and exert pressure at an international level to stop the rhinoceros horn trade.

Other threats to the survival of the species come from agricultural development and deforestation, which make the rainy season floods more disastrous every year, with the consequent further impoverishment of the natural environment.

Since the surviving rhinos are concentrated in limited areas, danger of epidemics or genetic weakening because of overpopulation in these areas exists; it would be necessary to transfer some of them to other suitable areas. An attempt in this sense was made in 1984 through 1985, by introducing into the National Park of Dudwa, in India, animals from Kaziranga and Chitawan. The operation has had a positive result, as seven of the nine animals transferred have survived and are in good health. This experience was also very useful for perfecting the techniques of capture and transport to be used in future reintroductions.

WILD BACTRIAN CAMEL

(Camelus bactrianus)

Ex	R	E	T	V	I	K

There are two species of the genus *Camelus*: the camel (*Camelus bactrianus*) and the dromedary (*Camelus dromedarius*).

Of the camel, two subspecies are recognized: one was initially described by Linnaeus in 1758 (*Camelus bactrianus bactrianus*), native to the former Soviet Union. The other is the *Camelus bactrianus ferus*, discovered and described by the Russian naturalist explorer Nikolay Przhevalsky. The latter seems to represent the wild form of the camel, whereas the former is the domestic form. In the domestic form, the two humps typical of this animal are larger than in the wild form, and the fur of the mantle is of a changeable color, and thicker. In the wild camel the fur is short, and the mantle is of a brown color and more uniform. The wild form survives in some arid areas of central Asia, but, even though protectionist measures have been taken, it seems to be on the verge of extinction now.

These animals live in rather arid regions, where they feed on grass and shrubs. The camels are the largest of the Artiodactylae, varying from a little more than 200 cm

(78 in) to about 350 cm (136 in) in length: they are 230 cm (90 in) in height and weigh 500 kg (1100 lb) or more. The neck is very long, and special glands in the occipital region of the head secrete an oily liquid. The domestic camel has a typical callosity on the joints, on the stomach, and on the parts where the limbs fit into the main skeleton, which the fetus also has. The wild camel does not have this callosity. Domestic camels present typical adaptations to life in the desert, with precise mechanisms to regulate body temperature in relation to the necessity of saving water.

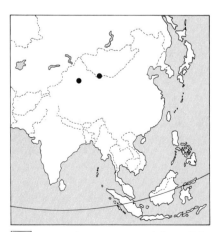

● Distribution area of the wild Batrian Camel.

CALAMIAN DEER
(Axis calamianensis)

Ex	R	E	T	V	I	K

The Calamian deer is of medium size, standing about 80 cm (31 in) high at the shoulders and weighing about 80 kg (176 lb). It differs in appearance from the other species of the genus *Axis* by its short muzzle and short, rounded ears. Its coat also has certain features that set it apart. It has white spots on the chin, that, unlike those of other species, extend toward the neck to form a large mark that includes the throat. The front part of this mark is divided in two by a narrow, tawny streak. The inner side of the base of the ears is white, the legs are dark brown, while the back is golden brown. The rump patch is orange. The antlers, which are lyre-shaped, are a reddish-brown, and the main bar bends first toward the back and then toward the outside. Normally, the length of the antlers reaches 75 cm (29 in), but the longest are up to 1 m (39 in). The usual number of points is six, but some antlers have eight or more. The antlers of most of the adults fall off between August and September; they then completely

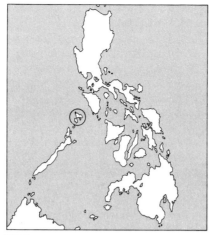

Distribution area of the Calamian deer

regrow but are still covered with velvet until March.

The mating season arrives usually about April or May. During this period the males become very possessive toward the females and engage in furious battles among themselves. Some animals receive serious, even fatal wounds from the antler blows. Most of the offspring are born between October and November, but the females may give birth during the entire year, and there is no precise mating season for all deer. Generally, the Calamian deer is a prolific animal, and the females may give birth to two or three young at once.

These deer have solitary habits, but it is sometimes possible to encounter small groups of them eating during the night, grazing on the grassy fields at the edges of the woods in which they prefer to spend the day, well-protected by the abundant vegetation. When frightened, the Calamian deer, like other members of the *Axis* genus, runs away with a very unique movement that some observers have liken-ed to the running of a pig.

These deer live in the Calamian Islands, in the Philippines. In 1975, there were about 900 specimens, but the reduction of this population began as early as World War II because of hunting. Since 1967, the authorities have forbidden hunting these deer, and possession of guns has been forbidden for the island inhabitants since 1972.

Another species of *Axis* is also threatened with extinction, the Kuhl's deer *(Axis kuhli)*, which lives on Bawean Island in the Java Sea. The numerical consistency of this species amounted to 500 specimens in 1969, but went down to about 200 to 400 in 1978. This reduction is a result, principally, of destruction of the natural habitat.

The Calamian deer lives on the Calamian Islands in the Philippines (map above). Because of hunting, its population is in decline and numbered in 1975 fewer than 1000 specimens.

NILGIRI TAHR

(Hemitragus hylocrius)

Ex	R	E	T	V	I	K

The Nilgiri tahr *(Hemitragus hylocrius)* is an Asian member of the Bovidae with an unusual appearance. Because of its morphological features, it looks like something between a goat and a sheep, and, in fact, the scientific name of the genus *(Hemitragus)* is Latin for "half-goat."

In this species, adult males are about 100 to 107 cm (39 to 42 in) high at the withers, with a weight of about 105 kg (231 lb). Females are considerably smaller, especially as regards their weight, which seldom reaches 30 kg (66 lb), whereas their height at the withers reaches 85 cm (33 in). The body color is brown tending to yellowish, with a darker color with large bright spots on the sides. Both females and calves differ in color from males, tending more to grayish. Both sexes have a long mane on the neck that covers also the shoulders and the upper part of the front legs. These animals always have a beard under the chin. The hair on the rest of the body is short and bristly. During the monsoon season, thin woolly hair is added to this covering, providing more protection against severe cold. The tail is short and of the same color as the body. The horns, flat at the sides, are very close together at the base, growing then wide apart and curving backward. They may reach a length of 40 cm (16 in) for males and 7 to 8 cm (2 3/4 to 3 in) for females. The females have only one pair of udders; this feature makes it possible to differentiate this species from the others of the same genus.

In the habitat in which the *Hemi-tragus hylocrius* lives, monsoons blow for 6 months of the year. These animals live preferably on peaks and near precipices, at altitudes between 1200 and 1600 m (3900 to 5200 ft), not far from wooded areas with open spaces for pasture in the middle.

At one time, this species was probably widespread in a large part of the Indian peninsula and reached northward as far as the Himalayas, occupying a wide area subsequently invaded by a similar species of larger size *(Hemitragus jemlahicus)*. At the moment, the Nilgiri tahr lives in the mountain areas of southern India, in the states of Kerala and Tamil Nader, populating the hills of Anaimalai, Nilgiri, Palani, and other mountain areas of the East Ghati.

The reasons for the decrease of their populations may be ascribed to such phenomena as intensive cultivations, breeding of domestic animals (mainly bovines), and the heavy deforestation carried out to enlarge pasture lands. Obviously, poaching is just as harmful and has been practiced mainly during the last 40 years for the skins and horns of these animals. All these factors have led to a complete disappearance of this species in many areas where it formerly lived.

An electricity company has organized special armed groups in the Nilgiri area to protect workers against possible attacks from dangerous animals; very often these gunmen shoot other animals, never incurring established penalties. It appears that the Nilgiri tahrs in the wild are not in danger of falling prey to other animals, and it seems also that they have never been infected by diseases that could have endangered their survival.

According to a census taken at the beginning of the 1970s, there were about 1500 to 2000 specimens of *Hemitragus hylocrius*, 800 to 1000 of which were living concentrated in two small areas, about 400 of them on the Nilgiri hills and about 100 on the Palani and Highwavy hills. In these areas, the situation appears to be good enough, considering the high number of young individuals inside the herds. Unfortunately, the same cannot be said of the state of Tamil Nadu, where the number has decreased from 1000 to 2000 head in 1950 to fewer than 60 in 1973.

This species is protected by law in all the areas in which it lives, but poaching is still a problem that cannot be easily solved. Four protected areas have been created for these animals in the regions where

Distribution area of the Nilgiri tahr

they live, the most important of which is Eravikulam-Rajamallay-Anaimudi, created in 1975. Of some importance were also some private associations that regulated hunting in some special areas of the region (the most famous was the Palani Hills Game Association). Unfortunately, the majority of these associations went bankrupt, and, consequently, poaching started again in these areas with great harm to animals. Plans for the protection of this species concern mainly poaching control and limitations to the expansion of plantations and cattle breeding, to avoid destruction of areas suitable for the survival of *Hemitragus hylocrius*.

It would also be of great advantage to reestablish private associations that, as in the past, could control the number of animals killed each year. It would also be important to consider the possibility of reintroducing the Nilgiri tahrs into those areas where they have been wiped out by man.

The *Hemitragus hylocrius* can easily be tamed and can easily adjust to life in captivity. There are some specimens in the Memphis zoo (Tennessee), in Trichur, in the Kerala District, and in the Trivandrum zoo, in southern India.

The causes for the decline of the Nilgiri tahr include intensive cultivation and breeding of domestic animals.

ORANGUTAN

(Pongo pygmaeus)

Ex	R	E	T	V	I	K

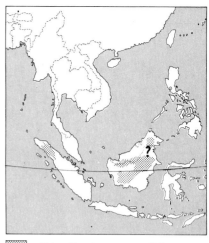

Distribution area of the orangutan

? Presence of orangutan in area

The orangutan is a robust ape; the male can reach 180 cm (70 in) in height when in an upright position, whereas the female usually reaches only 120 cm (47 in). The animal's back is structured in such a way as to be curved, while the belly shows marked prominence. The head is high, the ears small and without lobes, and the eyes are small and set close together. The muzzle is protruding, with a depressed nasal region, and the cheeks in the males have particular fatty protuberances, either bald or covered with fine hair. Both the upper and lower jaw are powerful, and in the subspecies that inhabits Borneo the face usually has little hair, whereas it is surrounded by a thick beard in the subspecies from the island of Sumatra. The older animals have clearly visible laryngeal sacs, with skin of a blackish-gray color. The forelimbs are particularly long and can reach as far as the level of the ankles when the animal is in an upright position. The span of the arms is considerable, and can reach up to 2.25 m (7 ft) in width. Both the thumbs and big toes are reduced and opposable. The rear limbs, on the other hand, are short and bandy. The whole of the body is covered with reddish fur, not particularly thick but quite long. It is extremely scarce on the breast, but abundant on the shoulders flanks, and limbs. All those parts of the animal that are not covered with fur are of a brownish-flesh or slate-gray color, with the particularity that in older individuals of the male sex they take on bluish tones. The color of the coat tends to grow darker with age.

The male and female orangutan live together only during the reproductive period and while raising the young.

On average the weight of an adult individual is in the region of 100 kg (220 lb), but cases of old males weighing over 150 kg (330 lb) have been recorded. The species seems to prefer swampy, level forests, but it has also been observed in hilly areas and at altitudes of over 1600 m (5200 ft) above sea level, according to the conformation of the territory in the various regions it inhabits.

The orangutan is purely arboreal and lives in primary and second-growth forests, preferring the higher layers of vegetation at heights that vary from 20 to 30 m (66 to 98 ft), although in the wild it has been observed on lower branches, occasionally even descending to the ground. The orangutan generally builds a sort of nest in the treetops (more rarely on the ground), using twigs and leaves, which it also uses often to cover its body and head in order to hide from strangers. Each day, every animal builds itself a new nest. At the age of just three years these animals start to build their own nest next to that of their mother, and up to the age of 4 to 6 years they continue to build in the vicinity of their parent.

The orangutan's diet is prevalently fruit-based, but this animal does not seem to disdain some types of flowers, berries, seeds, shoots, and birds' eggs. Both hands and feet are used during feeding to carry food to the mouth, and, holding on to branches with their hands, they sometimes pick leaves

Above, a mother orangutan and a young individual. The orangutan usually builds a sort of nest of branches and leaves in a tree. At about three years of age, the young orangutan builds its own nest near that of its mother and stays near its mother until four to six years of age.

or fruit directly with the mouth. The *Pongo pygmaeus'* day alternates moments dedicated to rest with others in which the animal feeds.

No predators of the orangutan are known, with the exception of man. This ape moves around in the trees in various manners, climbing, walking along the branches on all fours, and hanging from the branches from all four limbs, or with the upper limbs only. They have often been observed to move using brachiation (swinging by the arms), to walk in an upright position, and to walk with the knuckles of their hands resting on the ground. In spite of the variety of their methods of locomotion, the

orangutan is not particularly agile, and its movements are slow and deliberate.

The orangutan has the reputation of being a lazy and indolent animal, but it is perhaps better to say that it is extremely prudent, and every time it moves from one branch to another it avoids jumping, preferring to make certain of the strength of the new support and than move slowly onto it. The *Pongo pygmaeus* is a very silent ape, and it is not easy to note its presence in the forest. The only sound it makes a sort of grunt, produced by smacking its lips, to communicate with the other members of the group.

The orangutan cannot be considered a gregarious animal; isolated individuals, or mothers with their child, have been observed. The males usually spend the whole of their lives in a solitary state. Groups formed of a number of females with their young have also been seen, but never of more than six in number. Male and female couples form solely for the period

of time strictly necessary for mating.

The period of gestation is around 235 to 265 days, after which a single offspring weighing on average 1.7 kg (3 3/4 lb) is born. The newborn is suckled every 3 or 4 hours, for not more than 5 minutes at a time, for 4 months. After this period the milk is supplemented by food prechewed by the mother. Weaning takes place at around 3 years of age, at which time the young animal starts an independent life, although it still remains close to its mother. Sexual maturity in both sexes is reached at about 6 to 7 years of age. The interval between one pregnancy and another is rather long, and it has been calculated that, considering 50 years to be the average life span of each individual, a female does not give birth to more than 5 or 6 young during her life.

The orangutan is not a particularly early riser, and its day actually begins when the sun is already well up. It then abandons the nest in which it spent the night and moves off to consume its first meal of fruit and leaves. The morning passes with the animal still in the vicinity of the nest in which it spent the night, and when the sun reaches its zenith the orangutan carefully chooses the place in which it will build a new nest for the coming night. Thanks to its somewhat lazy character, it usually spends the early hours of the afternoon taking a nap. In the afternoon this species usually alternates periods of rest and periods in which it actively hunts for food. Before dusk falls, the orangutan carefully constructs its nest in the tree chosen in which to pass the night. Finally, before settling down to its nighttime sleep, it satisfies its alimentary necessities by eating a last meal.

Two subspecies of *Pongo pygmaeus* exist, and although they inhabit different regions they are not substantially different from one another. *P. p. pygmaeus* inhabits the forests of Borneo, whereas *P. p. abelii* can be found on the island of Sumatra, and differs from the former only in the slightly lighter coloring of its fur. *P. p. abelii* is widespread in the northern provinces of the island and in the vicinity of Atjeh. It is also present along the northwestern tract of the Wampu River. Again on the island of Sumatra, the subspecies habitually frequents the limits of the Gunung Leuser Reserve and the level forests of the West Langkot Reserve.

The exact distribuion of *P. p. pygmaeus* is not known. It is known that in Borneo this animal frequents the forest areas to the north of the Kapuas River, the central area of this region, and to the northeast a zone lying between the Mahakam and Kayan rivers. It seems that the whole area of forest lying in the southern part of Borneo, between the Barito and Mahakam rivers, is totally ignored by the orangutan. In the state of Sarawak (northern Borneo) it can be found mainly in the eastern regions, where small colonies of this subspecies can generally be found in the protected forests, near the eastern plains along the coast. In the northern area, the subspecies can be found in the Mount Kinabalu National Park. It seems that, in recent years, the orangutan has tended to concentrate in the level forests.

Apart from being hunted for its meat, the orangutan is also killed in some areas during the celebration of certain sacred rites, for the preparations of medicines, and also because it has the reputation of being dangerous to humans.

The process of numerical decline of the species is inexorable and continues each year. Once again it is humans who are to blame for this situation, with his expansion into once-virgin forests and his destruction of the orangutan's habitat to furnish in- dustries with timber, he is decimating and confining this species to extremely reduced areas. A further aggravant to this situation is the orangutan's low propensity for movement and adaptation to new areas and habitat. For this reason, many small colonies have been destroyed by the advance of civilization. It has been calculated that the global population of *Pongo pygmaeus* decreases annually by about 5000 individuals.

Numerical estimates of the population of the species carried out in 1984 revealed the presence of around 150,000 animals. The species is totally protected by law in its area of distribution, especially in the Gunung Leuser Reserve on Sumatra (which shelters 30 percent of the total number of living animals), in the Sarawak and Sabah state reserves, in Borneo, and also in other small protected areas. It is far more difficult to enforce the laws in isolated and poorly controlled areas.

The *Pongo pygmaeus* is listed in Appendix I of the Washington Convention, and trade of the species is therefore regulated by well-defined rules that are fully accepted by the nations signing this agreement, and severe punishment is provided for transgressors. Furthermore, thanks to this extremely important convention, any type of trade for profit is totally prohibited. Many scientists have studied the orangutan, evaluating and accurately comparing aspects of this species' behavior in the wild and in captivity. Thanks to these studies, to the undertaking of many people, and perhaps to the particularly awakened sense of respect for nature and for living things, programs have been started with the object of reintroducing and rehabilitating young orphan orangutans into protected areas in Borneo and Sumatra. Today, many zoos, understanding the gravity of the situation, have decided not to buy specimens of orangutan to show to the public, unless said specimens come through absolutely legal channels of importation, complete with all the certificates necessary for export.

In order to further protect the last remaining live animals it is necessary that the authorities of the countries in which the orangutan is still to be found include some areas of level forest (which today seems to be most commonly frequented by the species) in protected areas as soon as possible. It is also fundamental that nature reserves, or at least protected areas, be founded as soon as possible in Borneo.

The species is undergoing a constant decline in its numbers, losing about 5000 specimens each year from a total population estimated at 150,000 in 1984.

PHILIPPINE TARSIER

(Tarsius syrichta)

Ex	R	E	T	V	I	K

The tarsiers are small, unusual animals, nocturnal in nature and able to rotate their head completely around. This strange characteristic, together with their disproportionately large eyes, has created superstitions and beliefs among the native Southeast Asian populations. On a few islands it is believed that these creatures are evil spirits of the forest, and the Deiachi, famous Bornean head-hunters, believed it was an omen of bad luck to encounter a tarsier during a raid.

The Philippine tarsier was discovered in Luzon around the end of the 18th century by the Jesuit Father Camelli, who described it as "a small, long-tailed monkey." Uncertainties regarding the nature of this animal continued at length, to the point that Linnaeus, after having classified it as a monkey, inserted it among the opossums. Buffon considered it to be a jumping mouse species. It was only in 1777 that Erxleben placed the tarsiers close to the Lemuridae, thus considering them prosimians. Since the 19th century, the various tarsier species have been raised in captivity, and their morphological characteristics and behavior have been studied more carefully. Studies in the wild are very rare.

The Philippine tarsier is 15 cm (6 in) long, with a 20 to 25 cm (8 to 10 in) tail, and weighs about 1.5 hectograms (5 oz). Although it is small, this tarsier is the largest species of the three living today. Its slim body has short front limbs and very long rear limbs due to the extreme length of the tarsal bones. The long digits end with wide, round fingertips that facilitate grasping and clinging to tree branches. The roundish head has an unprominent snout and is connected to a quite small neck. The ears are slim and membranous, and the eyes are very large to the point that the facial portion of the skull is mainly occupied by the orbital cavities. The long tail is completely hairless, and it is only in the Celebes ghost tarsier (Tarsius spectrum) that the tail finishes with a little tuft of bristly hairs. The coat is grayish brown or dark brown, and on the hairless areas the skin has a slate gray color.

At the moment, the Philippine tarsier lives only on four islands in the southern archipelago (Samar, Leyte, Borhol, and Mindanao), populating the tropical rain forest with a striking preference for shrub vegetation and bamboo thickets. In a few areas, however, this little animal has also adapted to agricultural areas, frequenting cornfields and various types of plantations. It usually stations itself in the low vegetation area, between trees and shrubs, no more than 2 m (6 1/2 ft) from the ground. It is extremely agile, and thanks to its long, strong rear limbs it is able to complete acrobatic leaps from one branch to another, covering distances of 1 to 2 m (3 to 6 1/2 ft). While jumping downward it can cover up to 6 m (20 ft); on the ground it jumps using its rear legs.

Nocturnal in nature, this tarsier is mainly active after sunset and in the hours that precede sunrise. During the day it sleeps, attached to slender branches in the shadiest areas, sheltered from the light. It only rarely uses tree cavities. The tarsiers are generally not very sociable, living alone or as a couple. There is also very little contact between males and females. The Philippine tarsier nonetheless demonstrates great tolerance compared to other species of Tarsius and also forms small social groups. In this species physical contact is frequent, and during courting reciprocal grooming is quite common, mainly practiced by the males as compared to the females.

The diet of the tarsiers is prevalently insects, but these animals also hunt small birds, lizards, and

In numerical decline because of the destruction of its habitat, the rare Philippine tarsier is an arboreal rain forest animal. The numerical size of its populations is not known.

Distribution area of the Philippine tarsier

crustaceans. Their prey is usually captured with their hands and then bitten to death.

The reproductive behavior is not limited to a precise time of year, even if it seems to occur most frequently from August to December. After a gestation period of about 6 months, the female

The remarkably large eyes that are a characteristic of this small animal reveal its nocturnal habits. The pupils expand to surprising dimensions in scarce light.

gives birth to a sole offspring that weighs about 25 g (1 oz) and is already well developed, with open eyes and a moderate covering of fur. The little one remains attached to the mother's belly; however, on a few occasions the mother can also carry the newborn in its mouth. Tarsiers live for about 12 to 20 years.

The Philippine tarsier is considered relatively rare and on the decline, but no precise estimates of the species' numerical population exist. Easily domesticated, these tarsiers are frequently captured for commercial reasons, although after their inclusion in Appendix II of the Washington Convention this ended.

A greater problem is the destruction of their natural habitat, which has caused a drastic decline in the species, and if adequate measures are not taken, in the future this factor will continue to represent the major threat to their survival. This tarsier is also included on the Philippines' list of rare species and those in danger of extinction. It is hoped that inclusion on the list will prohibit the sale and exportation of these animals.

PROBOSCIS MONKEY

(Nasalis larvatus)

Ex	R	E	T	V	I	K

Borneo is a large island of 73,600 square km (28,400 square mi) lying in the Malaysian archipelago in the equatorial band. The climatic conditions are therefore extremely uniform and characterized by hot sunshine and abundant rainfall during the whole of the year. In this climate, the equatorial rain forest grows thick and luxuriant, accompanied, in the coastal areas and marshy regions around river estuaries, by the characteristic growth of mangroves. It is there, in the coastal mangrove forests and the rain forests along the river courses of Borneo that one of the strangest and most interesting of the monkeys lives: the so-called proboscis monkey.

The most obvious characteristic of this strange animal, as its name indicates, is its extremely long nose, which in adult males can reach the considerable length of 10 cm (4 in), hanging down like a small trunk to the level of the mouth. In the females and young animals the nose is shorter, although still of notable dimensions, and turned up. The large nose of the adult males appears to act as a resonance box and is related to the sonorous and powerful vocal emissions of this animal.

The proboscis monkey is a medium-large and robust monkey, and the males can reach a length of 66 to 76 cm (26 to 30 in) and a weight of not fewer than 16 to 22 kg (35 to 48 lb), whereas the females are much smaller, reaching lengths of 53 to 61 cm (21 to 24 in) and a weight of only 7 to 11 kg (15 to 24 lb). The tail is as long as the body or slightly longer. The coat is of a brick-red or chestnut color on the back, with an orange

The distinguishing characteristic of this species is its long nose (particularly in adult males). In its natural habitat, the proboscis monkey is an agile leaper.

area on the shoulders, and blends into grayish-brown on the lower parts of the body and on the limbs. A sort of cap of dark red fur is present on the head and extends down to the rear. The facial skin is meat-colored and embellished, on the cheeks, with thick hair of a creamy color.

The distribution of the proboscis monkey is limited to the island of Borneo, both in the districts of Sabah and Sarawak (Malaysia), and in Brunei and the larger central-southern region of Kalimantan (Indonesia). Their presence has also been observed on several small islands not far from the coast (Berhala, Sebatik). These monkeys live in groups of various sizes, ranging from a mini-

The proboscis monkey lives in coastal and mangrove forests and in the rain forests along rivers on the island of Borneo. The precise numbers of their populations are not known.

Distribution area of the proboscis monkey

mum of around 10 animals to 40 to 50 individuals, with extreme cases observed of about 100 monkeys.

Prevalently arboreal, they all descend quite often to the ground and are active during the daylight hours, whereas at night they retire to sleep among the branches of the mangroves. They are good swimmers, and one animal has been seen swimming underwater for almost 30 seconds. Almost exclusively vegetarian, they feed prevalently on the young leaves of the mangroves (*Rhizophora, Sonneratia*) and also on the fruit and flowers of various plant varieties (for example of the palm *Nipa fruticans*).

The little information obtained on their reproductive behavior is based on animals reared in captivity. The first birth in captivity took place in 1965 at the San Diego Zoo, in California. The young monkey, characterized at birth by bluish skin on the face, saw the light after approximately 5 1/2 months of gestation.

In Borneo, proboscis monkeys were hunted with slings and poisoned arrows by the Daiachi (natives of the island) for their tasty flesh, but this is certainly not the principal cause of the gradual decline of this species. Until 10 or 20 years ago the wood of the mangrove trees was considered of low economic value. Later, however, the use of these plants for the production of cellulose, tannins, charcoal, and construction material began, and concessions were issued even to foreign companies,

with the consequent deforestation of large stretches of forest. The situation is not yet serious, but there are fears for the future.

There is no estimate of the total number of proboscis monkeys living in Borneo, and relatively recent data indicate a population of 3000 animals in the area of Sabah and of 6400 in that of Sarawak.

The proboscis monkey is included in Appendix I of the Washington Convention and is also protected by the law in Sabah, Sarawak, and Kalimantan. Fortunately, they can be found in various parks and reserves, such as the Forest Reserve of Kabili Sepilok and the Klias National Park (in Sabah), in the Bako National Park, and the Gunung Pueh Forest Reserve (in Sarawak), not to mention the Kutai Reserve (in Kalimantan). It is also hoped that other reserves (such as the proposed Kayan Delta Nature Reserve, in eastern Kalimantan) will be founded, and that the species will be legally protected in Brunei.

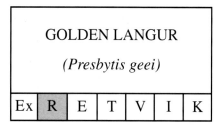

GOLDEN LANGUR

(*Presbytis geei*)

| Ex | R | E | T | V | I | K |

The langurs, or "leaf monkeys," are Asiatic monkeys with slender and slim bodies that are purely arboreal and almost exclusively phylophagous (that is, leaf-eating). They inhabit, in the form of 14 species, southern Asia from India to Indonesia and are related to the African colobus monkeys, belonging to the same subfamily (Colobinae). They have a long tail, are without cheek pouches, and, like all the leaf-eating monkeys, have a large, sacculated stomach with slow and complex digestive processes. The golden langur, which was previously considered to be a subspecies of the tufted langur (*Presbytis pileatus*), has a length of 64 to 72 cm (25 to 28 in) in the males and 48 to 61 cm (19 to 24 in) in the females. Its tail is

Distribution area of the golden langur, which extends from Bhutan to the borders of Assam in India.

long, varying from 70 to 80 cm (27 to 31 in) in the females to 78 to 94 cm (30 to 37 in) in males. Its coat varies in color according to the season, being yellowish-white during the summer and yellowish-brown in winter. The flanks are darker, almost reddish, while the ventral parts are lighter. A strong contrast is shown by the black face, decorated with a long, pale beard.

The area of distribution of the golden langur extends from central and eastern Bhutan to the borders with Assam (India), especially between the Sankosh and Manas rivers. These monkeys inhabit mixed deciduous forests and evergreen forests with thick bushy undergrowth up to an altitude of around 2400 m (7900 ft).

They live in groups of 10 to 20 individuals, usually guided by an adult male, and each group occupies an area of approximately 600 hectares (1482 acres). All the same, they do not show rigid territorial behavior, and the areas of different groups can overlap greatly. Relations between groups are passive, and aggressive behavior is hardly ever seen, even toward groups of other primates (sacred langurs or macaques). Timid and elusive, these langurs avoid meeting humans if possible, and when the group is molested, the females and young animals group together around the adult males, which emit repeated low-pitched noises.

Purely arboreal, the golden langur starts moving around just after dawn and remains active throughout the day. It feeds mainly in the morning and late afternoon, resting for a couple of hours around midday. It eats leaves, flowers, and shoots of various species (especially cardamom and kadoda figs) and descends to the ground only to drink, going to the banks of streams in groups.

No precise estimate of the numbers of this animal exists, but it is thought that the species is not in immediate danger of extinction, as it is relatively numerous at least in Bhutan. The area of distribution, however, is rather restricted, and this fact could prove to be very harmful in a case of sudden lack of balance in the natural habitat. Fortunately, this species, which is listed in Appendix I of the Washington Convention, is totally protected in India, where it can also be found in a nature reserve in Assam (Manas Sanctuary). In Bhutan the area around the Manas River has been declared a protected area by the government, and this langur, discovered in 1953 and immediately judged of particular beauty, should not be running any imminent risks.

Below, a group of langurs moving across their natural habitat in Bhutan. The species, identified in 1953 and protected, does not seem to be in imminent danger. Right, the coat of the golden langur changes with the season, being a yellowish white during the summer and yellowish brown in winter. Its face is black.

The macaques are robust, thick-set monkeys, with a distinct supra-orbital crest and, with few exceptions, a relatively short tail that is sometimes even reduced to a small nodule

Extremely adaptable animals, macaques inhabit both forest areas and open or semirocky areas, easily getting used to the presence of humans and sometimes even living in the vicinity of inhabited areas (for example *Macaca mulatta*). Macaques are widespread in southern and eastern Asia, from India to Japan, with the exception of the Barbary macaque *(Macaca sylvanus)*, which lives in North Africa.

Of the 16 living species of macaque, the majority have wide areas of distribution and relatively numerous populations, but several species have suffered a rapid numerical decline during this century and are seriously threatened with extinction. This is the case, sadly, of the so-called lion-tailed macaque *(Macaca silenus)*, a rare species that lives exclusively in southwestern India. Because of the large mane surrounding its head, this macaque is also known as the lion monkey (but this could cause confusion with the "lion monkeys" of South America belonging to the family Cebidae). The name *lion-tailed macaque* refers to the tuft of hair at the end of its tail. Another name often given this animal *(uanderù)* seems to derive from a linguistic generalization, as it is the Singhalese term used in Sri Lanka (Ceylon) to indicate langurs.

The size of the lion-tailed macaque is slightly smaller than that of the other macaques. Its weight varies between 7 and 9 kg (15 and 20 lb), whereas the length of its body is around 51 to 61 cm (20 to 24 in) in the males and approximately 46 cm (18 in) in the females. The tail is relatively short—25 to 38 cm (10 to 15 in)—

Distribution area of the lion-tailed macaque in the western Ghati Mountains at the southernmost end of the Indian peninsula.

and terminates in a small tuft of hair. The coat in both sexes is of a beautiful shining black all over the body, except for the area around the face, which is surrounded by a wide collar of long, grayish hair.

This macaque lives in the western Ghati Mountains, in the Indian peninsula, but whereas at one time its area of distribution extended as far north as Goa (state of Karnataka), today it appears to be restricted to a few hilly areas in the state of Kerala and at the border with the state of Tamil Nadu (the hills of Nilgiri, Anaimalai, Ashambu, and perhaps Cardamom).

Unlike the other macaques, it is exclusively arboreal and prefers evergreen rain forests. In these surroundings, which it shares with the Nilgiri langur *(Presbitys johnii)* and the bonnet macaque *(Macaca radiata)*, the lion-tailed macaque is found mainly at altitudes of between 600 and 1100 m (1970 to 3600 ft), although it can climb to higher altitudes or descend lower, even down to 150 m (492 ft) in the band of humid deciduous forest.

It is not easy to observe these primates in their natural habitat because of their timid nature, their tendency to avoid meeting humans, and because of their dark coat and their habit of staying in the highest layers of the vegetation. For a long time the principal observations on the biology of these animals were therefore carried out in zoos, in

several of which the lion-tailed macaque breed successfully (for example at New Delhi). But in 1973 research in the wild was started and new information was added that helped explain various aspects of this animal's behavior.

The lion-tailed macaque lives in groups of around 15 individuals, with rather wide group—up to 5 square km (2 square mi)—which usually overlap. It feeds mainly on fruit, but also eats flowers, leaves, herbaceous plants, and seeds, not to mention larvae and insects, which it hunts actively under the bark of trees or among rotting trunks, and occasionally small vertebrates (lizards and tree toads). The groups can effect seasonal migrations according to the ripening of different types of fruit, but some tree species bear fruit during almost the whole year, such as the *Artocarpus* (breadfruit tree) or the *Cullenia exarillata*, and the presence of these trees is vital in the living area of these animals.

Rainfall is always abundant in these regions, and the lion-tailed macaque does not need to descend to the ground in order to drink, as it is sufficient for them to lick the leaves or drink the rainwater that gathers in the hollows of trunks or in forks in trees. However, in the western Ghati Mountains one or two periods of relative drought can occur during the year, and on these occasions the groups of macaques come down from the trees to drink at the edges of streams, which can always be found in their area of distribution.

Sexual maturity is reached at around 5 years of age in the female and at around 8 years in the male. At birth, the young have pale pink skin covered by short brown fur, but after a month both skin and fur begin to grow darker.

In captivity, a life span of 17 to 18 years has been witnessed. No exact estimate of the size of the populations exists, but this macaque is rare throughout its territory and is probably already extinct in the northernmost part (the state of Karnataka). Estimates, not entirely reliable, of the popula-

The lion-tailed macaque, with its characteristic large mane circling its head, is strictly arboreal.

34

tion of the species stood at 1000 specimens in the 1960s and only 500 animals in 1975, but it is too difficult to carry out an exact census.

More than hunting, carried out for food, or the capture of animals for commercial purposes, it has been the drastic reduction of its natural habitat that has brought this species of macaque to the brink of extinction. The destruction of the forests to make way for agriculture and tea and coffee plantations or for the cultivation of exotic species with rapid growth (eucalyptuses and acacias) has had a disastrous effect on the population of this macaque, which, since it dwells exclusively in the primary rain forests, adapts badly to second-growth forests and the artificial environments constructed by humans.

Studies carried out in the Ashambu hills have underscored the necessity of at least 130 square km (50 square mi) of intact and continuous forest to support a population of 500 macaques, the minimum number necessary to guarantee stability and continuity in time to the population itself. Intact areas of such a large size still exist in the southernmost parts of the western Ghati Mountains, but urgent measures are necessary to preserve them from economic exploitation.

The species can be found in the Periyar Sanctuary and, since the beginning of 1976, the state of Tamil Nadu has formed a protected area on the Ashambu hills (the Kalakkadu Forest Sanctuary) with the main object of protecting the lion-tailed macaque, but other protected areas must be founded, in collaboration with the state of Kerala (for example between the valley of the Bharani River and "Silent Valley").

The species is listed in Appendix I of the Washington Convention and has been protected in India since 1972 (India's Wildlife Protection Act), but this protection is not always effective. It is necessary for greater control to be obtained over poachers and animal traders, for access to the areas that are most important for the survival of these macaques to be limited, and for the felling of trees to be rigidly regulated, with no further concessions for the exploitation of timber in those areas.

Only by safeguarding their forest surroundings will it be possible to save the lion-tailed macaque, which, along with the rhinoceros, the snow leopard, and the Asian lion, is one of the rarest and most threatened species in India.

No precise information on the size of the population of this species is available (it was estimated at 500 specimens in 1975), but it is thought to be on the verge of extinction due to the reduction of its natural habitat.

DOUC LANGUR

(Pygathrix nemaeus)

Ex	R	E	T	V	I	K

The Douc langur differs in several morphological and skeletal characteristics from the other langurs (genus *Presbytis*) and seems closer to the snub-nosed monkeys (*Rhinopithecus*, recently included in the genus *Pygathrix*).

Information on this rare species is extremely scarce, due to the fact that its area of distribution is limited to areas of the Indochinese peninsula that have been, over past decades, the site of bloody warfare. Only in recent times has research been carried out in its natural habitat, and this research has, sadly, confirmed the conditions of extreme danger to the future survival of this species.

The Douc langur is a medium-large-size monkey and reaches a length of around 61 to 76 cm (24 to 30 in), plus a similar length of tail. The females are generally slightly smaller than the males. Its most striking feature is the color of its coat, which, in the contrast between dark and light parts, makes the animal look as though it were wearing a shirt and shorts. The general color is blackish-gray, but the forearms are white, and the legs, from the knee down, are chestnut. Hands and feet are completely black, and the tail is white, as is a triangular area at the base of the tail itself. White areas also extend onto the cheeks (long white side whiskers turned back) and on the throat, making a clear contrast with the collar of reddish-brown around the neck and the band of the same color across the forehead.

Two subspecies of Douc langur are known: *Pygathrix nemaeus nemaeus*, the typical form, and *P. n. nigripes*, which differs from the former in its coloring, which is darker overall. In the latter subspecies, in fact, the forearms are not white, and the rear limbs are of a uniform blackish color; furthermore, there is a black area

around the eyes, and the side-whiskers are shorter.

This langur inhabits the tropical rain forests of the Indochinese peninsula, with an area of distribution that extends from Laos to Vietnam and perhaps also into Kampuchea (Cambodia). Ancient records indicate its possible presence on the island of Hainan. Only since the 1970s has research begun in Vietnam, while other information on the biology of this

Information on the biology of the Douc langur comes for the most part from specimens raised in captivity in zoos.

species derives from observations made on specimens reared in captivity (for example, in the zoos of San Diego, California, and of Cologne, Basle, and Stuttgart in Europe).

Like all the langurs, the Douc

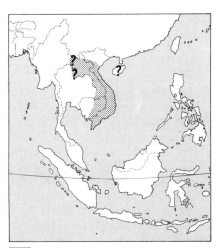

langur feeds almost exclusively on leaves, although it also eats the fruit of numerous plant varieties. Groups of this species have, in fact, been observed to move around within the limits of their territory according to the period in which various fruits ripen. The Douc langur eats by raising food to its mouth with its hands, rather than taking it directly with the lips, and it drinks little and rarely, obtaining most of the water necessary from the fruit and other plant substances on which it feeds.

The groups are formed of a number of individuals ranging from 4 to 15, but during the 1930s groups of even 40 or 50 specimens were seen. Movements from place to place occur mainly along well-defined tracks through the branches, using a normal bipedal form of locomotion, but adult animals sometimes reveal unexpected acrobatic ability, climbing up the trunks with extreme rapidity and even making leaps of great distance.

Little is known about the reproductive behavior, but in the wild it seems that most of the births take place between February and June, probably in relation to the seasonal availability of food and the better climatic conditions.

A Douc langur and offspring. This monkey inhabits tropical rain forests of the Indochinese peninsula. Its most numerous populations are believed to be in Laos.

In zoos it has been established that the gestation period lasts around 6 months. In its natural habitat, the Douc langur has very few enemies (one or two carnivores and snakes), and in natural conditions the only limiting factors for its colonies are the availability of food, sickness, and natural disasters.

Since the beginning of this century, however, it has been humans who have revealed themselves as a natural calamity for the species, causing both directly and indirectly a constant numerical decrease and bringing the species to the brink of extinction. Hunting carried out by local tribes and the destruction of the tropical forests for the exploitation of their wood are only partial causes of the decrease in numbers of these langurs. The principal damage has been done by the wars that have bloodied the region and everything this warfare brought about; destruction of the natural habitat by bombing and, above all, the use of chemical defoliants and napalm bombs, not to mention the direct killing of langurs as a form of target practice for soldiers and guerrillas.

At present, an exact estimate of the number of living animals does not exist, but it appears that the most numerous populations are those of Laos, from which almost all the specimens living in zoos originate. Only fragmentary information can be obtained on the subspecies *Pygathrix nemaeus nigripes*, and it is thought that some specimens may survive in the areas to the southwest of Saigon, near the border between Vietnam and Kampuchea.

In the 1960s, the species was protected by the government of Hanoi, but this legal protection had very little effect. It is now hoped that, with the reduction in fighting and the improvement of economic and alimentary conditions in the country, the pressure caused by hunting for food may be reduced. All the same, it would be necessary to establish specific reserves in southern Vietnam, central Vietnam (on the border with China), and in the region bordering Laos to create well-founded hopes for the survival of the species.

<table>
<tr><td colspan="7">NILGIRI LANGUR

(Presbytis johnii)</td></tr>
<tr><td>Ex</td><td>R</td><td>E</td><td>T</td><td>V</td><td>I</td><td>K</td></tr>
</table>

The Nilgiri langur, known as *karu-manthi* or *karum kurangi* in Tamil dialect, is, like all the langurs, a slender and agile animal, although of medium-large dimensions. It reaches a length of 78 cm (30 in) and a weight of 13 kg (27 lb), with an extremely long tail—approximately 70 to 95 cm (27 to 37 in). The females are slightly smaller and lighter, reaching a maximum length of around 58 cm (23 in) and a weight of 11 kg (24 lb). The coat is shiny black, with grayish-white fur on the back and at the base of the tail in adult animals. The females have a white spot on the internal surface of their thighs. The head is yellowish-brown, with darker areas on the top (brown) and on the throat (reddish-brown).

This langur lives in southwestern India, with a distribution limited to the hilly areas of the western Ghati, south of Coorg, and the largest numbers are found in the hills of Nilgiri, Anaimalai, Palni, Brahmagiri, and Ashambu. This species usually inhabits evergreen forests at altitudes of between 900 and 1200 m (2950 and 3900 ft), but it is extremely adaptable and can descend to lower levels—500 to 600 m (1640 to 1970 ft)—or climb higher, up to over 2000 m (6500 ft).

Its natural habitat is formed of extensive evergreen forests that occasionally form the so-called *sholas,* vegetation consisting of bush or strips of evergreen forest, surrounded by open areas of prairie and crossed by brooks or streams. The tree species can even reach 60 m (197 ft) in height, while at ground level a rich brushy undergrowth grows.

The Nilgiri langur lives in groups of varying size (between 10 and 30 individuals), generally with a single adult male having the functions of reproduction and protection of the territory. Early in the morning the male emits a loud cry to which the males of nearby groups respond, and soon the forest is ringing with their calls. In spite of this evident territoriality, meetings between groups are almost always pacific, and the males of each group limit themselves to gesturing and vocal exhibitions at the edges of their territory, often remaining at distances of up to 25 m (82 ft) from one another.

The size of their territory is extremely variable: on the hills of Ashambu an average size of 25 hectares (62 acres) has been observed, whereas on the Nilgiri

langur is rigorously forbidden, and a colony of this rare species is protected in a nature reserve in the region of Szechwan (Wanglang Nature Reserve).

It is necessary, however, to carry out further research, both in order to gather more information on the biology of this animal and to identify the strategic areas necessary for its survival in order to open specific nature reserves. This also goes for the *Rhinopithecus roxellanae bieti*, which has an extremely small area of distribution, and for the other species of snub-nosed monkey living today: the *Rhinopithecus brelichi* and the *Rhinopithecus avunculus*. The latter is the so-called Tonkin monkey, a distinct species of snub-nosed monkey, which is also rare, being found in the region of Tonkin, on the borders between China and Vietnam. It is slightly smaller in size than the other snub-nosed monkeys described, 56 to 62 cm (22 to 24 in) in length, plus 80 to 85 cm (31 to 33 in) of tail, and the color of its coat is slightly darker, with a blackish back and limbs and yellowish-white under-parts. Nothing is known about its living habits and the number of living animals, and further research in this direction is to be desired.

In China, the snub-nosed langur was hunted for its highly prized coat; only Manchu functionaries were permitted to wear them. Today, hunting is prohibited, and a group of the species is protected in a nature reserve.

<table>
<tr><td colspan="6">**BRELICH'S SNUB-NOSED LANGUR**

(Rhinopithecus brelichi)</td></tr>
<tr><td>Ex</td><td>R</td><td>E</td><td>T</td><td>V</td><td>I</td><td>K</td></tr>
</table>

BRELICH'S SNUB-NOSED LANGUR *(Rhinopithecus brelichi)*

Ex	R	E	T	V	I	K

PIG-TAILED LANGUR *(Simias concolor)*

Ex	R	E	T	V	I	K

The distribution area of the pig-tailed langur in the Mentawai archipelago off the coast of Sumatra.

Almost nothing is known about this strange and mysterious monkey, as, unlike its better-known and studied relations, its existence has been deduced exclusively from the discovery of two stuffed specimens from southwestern China. The first of these, furnished by a Chinese hunter, goes back to 1903, while the second skin is that of an animal killed in 1963 on the southeastern slopes of Mount Fang Chin, in the region of Kweichow.

From the description of these examples, it can be seen that this species is larger than the golden monkey, having a length of around 73 cm (28 in), plus about 97 cm (38 in) of tail. The color of its coat is yellowish-brown, with yellowish shading on the arms and a large off-white area on the shoulders; the head is of a light gray color, darker toward the crown. The tail is blackish, with a white tip.

This snub-nosed monkey also inhabits mountain forests, even at high altitudes and fairly low temperatures during the winter season. Its long fur—70 to 80 mm (2 3/4 to 3 in)—therefore represents an obvious adaptation to environmental and climatic conditions. It is thought that its area of distribution is rather restricted—perhaps around 5000 square km (1900 square mi)—and limited to the region of Kweichow.

The pig-tailed langur has long been considered one of the most rare and mysterious of the living primates. It was only in the 1970s that research carried out in several parts of its restricted area of distribution clarified many aspects of its behavior and distribution.

This species, which some experts consider to belong to the genus *Nasalis* due to its probable ancient relationship to the proboscis monkey, lives exclusively in the Mentawai archipelago, a group of small islands situated just off the southwestern coast of Sumatra (Indonesia). These islands lie in a particularly unstable part of the world, tormented by frequent volcanic eruptions, and their geological history is relatively short. First appearing as ridges parallel to the coast between the early and mid Pleistocene period (between approximately 1.5 million and 500,000 years ago), during the ice ages they remained connected intermittently to the mainland, due to the lowering of the ocean level. However, these small islands which together total an area of only about 7000 square km (2700 square mi), have now been completely isolated from Sumatra for at least 10,000 years, and the local fauna has evolved in a state of total isolation, especially those species that are incapable of crossing the wide arm of water lying between the archipelago and the mainland. It is not by chance, therefore, that 10 endemic species of mammal inhabit the Mentawai islands, among which are four monkeys, all rare and in danger of extinction. Apart from the pig-tailed langur, in fact, a species of langur (*Presbytis potenziani*), one of macaque (*Macaca pagensis*), and a variety of gibbon (*Hylobates klossi*) are all exclusive to these islands.

The pig-tailed langur is a monkey with a thick, stubby body, about 45 to 55 cm (18 to 21 in) in

length, with robust arms and elongated fingers, all related to its prevalently arboreal habits. Its tail is short—about 15 cm (6 in)—and hairless, with the exception of a tuft of hair at its extremity. Another characteristic is its very long, turned-up nose, which is also common to the other snub-nosed monkeys (genus *Rhinopithecus*) and the females of the proboscis monkey. Its coat is blackish-brown on the back, with cream-colored shading in the lower parts of the body. Mutants with lighter colored coats have been observed, attributed by some experts to a distinct, although probably unjustified, northern subspecies (*Simias concolor siberu*). The hands, feet, and face are black, and the cheeks have thick, turned-back side-whiskers. In this species no important morphological differences exist between male and female.

The distribution of Simias concolor is limited to four islands in the Mentawai archipelago: Siberut, the largest, Sipora, North Pagai (or Utara), and South Pagai (or Selantan). In these islands, the pig-tailed langur populates the luxuriant tropical rain forest, especially in hilly areas. It has rarely been observed in secondary growth, and never in the coastal mangrove

Right, the pig-tailed langur. The map above shows the distribution area of the species, limited to the Mentawai islands off the coast of Sumatra.

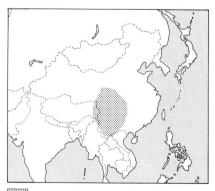

Distribution area of Brelich's snub-nosed langur

46

forests. This species is prevalently arboreal, but it also moves quite rapidly on the ground. It feeds almost exclusively on leaves and shoots, integrating its diet with berries or other fruit, and even with crabs and other small aquatic creatures.

The pig-tailed langur lives in small family groups of two to five individuals, formed by an adult couple with immature young. It does not demonstrate particular territoriality, and each family group occupies an area of about 25 to 30 hectares (62 to 74 acres), more or less overlapping that of other members of the same species. When two groups meet, the males come toward each other, exchanging a low vocal signal, returning afterward to their respective family group without showing hostility.

Unlike the proboscis monkeys,

the snub-nosed monkeys are extremely quiet, and the lack of signals and loud vocal sounds may represent an adaptation to the absence of large predators in their surroundings. The life in small family groups, on the other hand, may be a later adaptation, developed in order to escape with greater ease from preying by humans.

The natives of Mentawai have always hunted the monkeys (using slings and poisoned arrows) for food, but a complicated system of tribal habits and religious tabus kept these primitive populations in perfect equilibrium with their natural habitat until the start of the 20th century. Later, progress radically altered the culture of the local populations, and the exploitation of natural resources, a privilege granted also to foreign companies, has begun to eat away the delicate eco-

An arboreal monkey of tropical rain forests, the pig-tailed langur is threatened by the cutting down of the forest for timber and the expansion of agriculture.

logical equilibrium of the islands. The forest has made way for cultivations and plantations and the exploitation of timber has worsened the situation, accelerating the destruction of the primary forests, which are essential to the existence of the snub-nosed monkeys.

Monkey hunting, carried out with modern and more efficient methods, and without the traditional limits imposed by primitive tabus, has also contributed to the decline of the rare species present in the area, and the situation appears particularly critical on the island of Siberut, the largest and most thickly inhabited of the archi-

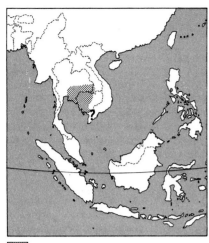

 Distribution area of the pileated gibbon

? Possible presence of species in area

remain with the groups until 6 to 7 years of age. When moving around, this species makes the most of its long, powerful arms to effect long leaps between the branches of trees. When on the ground, it assumes an erect position and moves easily, swinging its body from side to side and keeping its arms extended laterally in order to balance better. If it must run or move rapidly on the ground, it uses its long upper limbs as crutches, resting the back of its hands on the ground.

The *Hylobates pileatus* is distributed in various countries: in Laos, in Kampuchea (to the west of the Mekong River), and in the southeastern part of Thailand. Sadly, the species is already extinct in Vietnam. In Thailand the pileated gibbon can be found in the evergreen forests of the districts of Jurin, Buriran, and Prachinburi. In the southern part of the country it is certainly present, as it is in the west, in the Khao Yal National Park and the Khao Sol Dao Sanctuary. No information is available on the exact distribution of the species in Laos and Kampuchea.

The number of living animals is not known, but this number is certainly in continuous decline. In actual fact, several areas of forest are already empty of gibbons. In 1977, thanks to scientific research, it was established that between 1300 and 3000 specimens were still present in the Khao Yai National Park, and not more than

500 to 1500 individuals in the Khao Sol Dao Sanctuary. Both the national park and the sanctuary are the only areas in which the pileated gibbon is protected, and it is thought that elsewhere the number of animals is extremely low. The numerical situation of the species in Kampuchea is not known, but it is thought to be even worse than that in Thailand.

The main causes of the numerical decrease of this species is the brutal deforestation that, especially in Thailand, goes on at a rate of 5 to 10 percent per year. The local populations continue to fell tracts of the forest to provide space for cultivations, for the building of roads, and for the construction of dams. Aerial photography has shown that two thirds of the area of distribution of the pileated gibbon is being exploited by humans. Furthermore, the females of this species are killed for meat, and the young animals are captured and domesticated to be sold.

The *Hylobates pileatus* is listed in Appendix I of the Washington Convention, and the killing, possession, sale, and exportation of the species are forbidden in Thailand. Sadly, with the exception of the protected areas, poaching is frequent in all the other areas.

AMAMI RABBIT

(Pentalagus furnessi)

Ex	R	E	T	V	I	K

The amami rabbit was first described by Stone in 1900 on the basis of two specimens captured on Oshima Okinawa Island. This rabbit is 43 to 51 cm (17 to 20 in) long and weighs about 2.5 kg (5 1/2 lb). It has very small eyes, short ears—just 45 mm (1 3/4 in)—and a very short tail, not even 15 mm (1/2 in) long. Its rather stocky paws have very strong and long claws—20 mm (3/4 in)—rather unusual for a rabbit. Each front limb has five digits and each back, four. Its

Latin name, *Pentalagus*—based, in fact, on the Greek word *pente,* "five"—refers to the fact that the first examined specimens possessed only five molars per jaw, while all other Leporidae have six. Afterward, however, it was found that this characteristic is very rare in this species, too.

The amami rabbit is not especially skilled at running because its lumbar vertebra transversal processes, attached to the posterior adducent muscles, are short and wide, instead of being narrow and long as is the case in all other genera of this family. In addition, its tarsus and metatarsus are very short. Its fur is made up of a lead-gray soft undercoat and a long thick bristly main coat. The latter is dark brown on the back, reddish on the sides, and light brown on the lower sections. A black stripe runs down the back from the neck to the base of the tail.

The *Pentalagus furnessi* is native to the Ryukyu archipelago, a series of small islands located southwest of Japan between Kyushu and Taiwan. Today, this rabbit can be found only on two of these islands: Omami-Oshima (718 square km [277 square mi]) and Tokuno-Oshima (247 square km [95 square mi]). The climate of these islands is typical of the humid tropics, modified by the Kuroshio air current. It rains often and there are broadleaf and evergreen forests, including oak, palm, and bamboo trees. The amami rabbit lives in these forests on mountain slopes, always near waterways. Nocturnal in nature, it sometimes enters these dense forests, but generally prefers to remain at their borders, where the arboreal vegetation is thinner, permitting the growth of a grassy underbrush.

It usually lives in couples or in

The amami rabbit lives in forests on mountain slopes near waterways on two islands of the Ryukyu archipelago, an area with a humid tropical climate. The current available data give its population fewer than 1000 head.

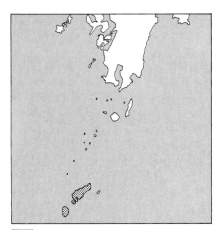

Distribution area of the amami rabbit on two islands

small family groups. The gestation period lasts about a month and a half and, depending upon when the mating occurs, the offspring can be born from August until January. Usually just one bunny is born, occasionally two, and it is kept in a separate burrow, far from that used by the parents. The burrow is dug in soft ground and can reached by a tunnel 1 to 2 m (3 to 6 1/2 ft) in length. The mother covers the burrow with leaves and fur and comes to nurse her offspring at night. During the day she never goes near her offspring, but during the first week she protects it from predators by covering the tunnel's entrance with soil. When the offspring is 40 days old it comes out of its refuge and is taught how to search for food. The diet consists of bamboo and banana sprouts, vines, fruit (pineapple and oranges), sweet potatoes, and—when they get bigger—bark.

The burrows of adult rabbits are located underground, under large rocks and at the foot of big trees. There is usually a tunnel, from 30 to 200 cm (12 to 78 in) in length, leading to the burrow. Every burrow has an additional tunnel that is used for ventilation and also as an emergency exit. This dwelling is covered by a thick layer of dry leaves and has a diameter of 20 to 150 cm (8 to 58 in), depending on the number of rabbits that live there.

One or two families of these rabbits can live in a valley. They take advantage of the waterways to mark their territorial boundaries, depositing excrement on dry water beds. These creatures love water and enjoy taking baths in currents and pools. When this animal is being chased by a predator it takes refuge in crevices between rocks and the cavities of hollow trees. From the latter it is able to reach its hole through the trunk of the tree.

It is not known how long the species live in the wild, but in captivity specimens have lived for as long as seven years. Whether due to its population or its area of distribution, this animal has decreased greatly in number in our century. Research reveals that there are about 1000 specimens still living, but not all sources consider the situation particularly tragic.

This rabbit is now endangered because it was not rigorously protected from hunting until 1921. In addition, deforestation has reduced its natural habitat. Another element that has contributed to its decline is the large presence of domesticated animals (dogs and cats) that have become wild. It is obvious that to protect these animals stronger hunting laws must be proposed and parks and reserves must be established in the areas where they are still present, limiting the impact of humans in this region.

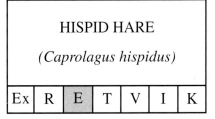

HISPID HARE

(Caprolagus hispidus)

Ex	R	E	T	V	I	K

The hispid hare is 47 cm (18 in) long, and its ears, tail, and paws are rather short compared to those of other Leporidae. Its fur is rather rough (as its Latin name suggests: hispidus means "bristly"), and the upper regions of its body are dark brown, while the more central areas are whitish.

This rabbit could originally be found in all the regions surrounding the Himalayas, from the Uttar Pradesh to western Assam. A few years ago it was believed to have become extinct except in the Manas Wildlife Sanctuary and in Assam's Forest Reserve. However, in 1984 these animals were discovered in southwestern Nepal, northeastern Uttar Pradesh, and on the prairies of Bengala.

This species' continued survival is endangered principally because its natural habitat, the prairies, are being destroyed. The tall grass that grows there is used by local farmers to construct roofs and is thus regularly cut down and burned. These actions continually reduce this species' habitat. In addition, hunting, although illegal, has contributed to their decline.

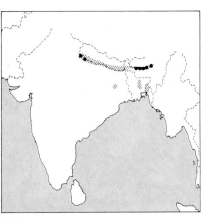

Distribution area of the hare

• Areas of recent reports of sightings

MINDANAO GYMNURE						
(Podogymnura truei)						
Ex	**R**	E	T	V	I	K

KITTI'S HOG-NOSED BAT						
(Craseonycteris thonglogyai)						
Ex	R	E	T	V	I	**K**

SINGAPORE ROUNDLEAF HORSESHOE BAT						
(Hipposideros ridleyi)						
Ex	R	**E**	T	V	I	K

The Mindanao gymnure is probably one of the last representatives of the primitive Erinaceidae. It is 13 to 15 cm (5 to 6 in) long with a slight hairy tail of 4 to 7 cm (1 1/2 to 2 3/4 in). It has an oblong snout, and its dentition is distinctive for its long and well-developed canines. The long, soft, thick hair is gray on the back and clear on the belly.

They live exclusively in the mountain forests of the island of Mindanao (Philippines), on Mount Apo and Mount McKinley in the south and on Mount Katangland in the north at altitudes between 1600 and 2300 m (5200 to 7500 ft).

We know little about its habits, but they are probably similar to those of the shrew. Its numerical population is not known. Included on the IUCN *Red List*, it is considered in danger of extinction.

The survival of this species is tied to the protection of its habitat, given that intense deforestation has already destroyed most forests below 1400 m (4600 ft). It is protected in the Mount Apo National Park, where above 500 m (1600 ft) the vegetation is still intact.

The Kitti's hog-nosed bat is both one of the smallest mammals—it is 3 to 3.3 cm (1 to 1 1/4 in) long and weighs just 2 g (0.07 oz)—and also one of the most recently discovered. The first sighting of this species occurred in October 1973 in the province of Kanchanaburi in central-western Thailand. Its dorsal coat is long and grayish brown in color, while on the belly it is shorter and paler in color.

During the day, this bat shelters in granite grottoes, where it hangs in the most hidden spots. It lives in small communities of 10 to 15, but each member maintains a certain distance from the others, and on the whole they do not seem very sociable. At dusk, it goes out to look for food, flying around the tops of bamboo plants and teak trees. It feeds primarily on insects it finds on leaves and is probably able to catch them in flight.

Although there are no estimates on the population, the species is believed to be rare. However, since the grottoes of Southeast Asia have not been thoroughly explored, one cannot exclude the possibility that this bat lives in other areas.

The Singapore roundleaf horseshoe bat is a unique-looking bat that has a very restricted distribution area on the Malay peninsula. The first specimen was discovered in 1910 in the Singapore Botanical Garden and for many years remained the only known example. This medium-sized bat has a forearm 4.7 to 4.9 (1 2/3 to 1 3/4 in) long. The color of its mantle is dark brown or black. The appearance of its head is very unusual. Its large, wide, pointy ears are subtriangular in shape and 2.1 cm (2/3 in) long. Their bases are covered by fur. The nasal layer is very large, almost completely covering the snout. It is subdivided into three rather different and bizarre parts, of which the middle one is the thickest. Principally in the males there is, behind the nasal layer, a small sac that secretes a waxy substance.

In 1975, 65 years after the original discovery of the Singapore horseshoe bat, a small population of 50 specimens was found in a peat forest, located 40 km (25 mi) northwest of Kuala Lumpur, in Malaysia. The strange

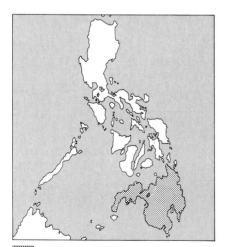

▨ *Distribution area of the Mindanao gymnure*

☐• *Distribution area of the Kitti's hog-nosed bat*

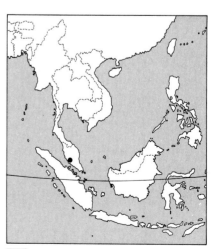

☐• *Distribution area of the Singapore roundleaf horseshoe bat*

thing about this discovery is that during the day the population takes shelter in a water pipe beneath a road. Very little is known about their behavior and biological makeup. It has been observed that while hanging, its ears and nasal layer contract.

Regarding reproduction, the birth of four offspring in April has been reported. The species is included on the IUCN *Red List* and is in danger of extinction. The principal threat to its survival is the razing of the peat forests, already rare and reduced to small, isolated areas.

BEIJI DOLPHIN

(Lipotes vexillifer)

Ex	R	E	T	V	I	K

The Beiji dolphin is a typical river dolphin characterized by a strong body and a long and very slender snout. The adults are about 2 to 2.5 m (6 1/2 to 8 ft) long, and the females are usually a little bit larger than the males.

At birth, they are about 80 cm (31 in) long. Their back and sides are grayish in color, while the underside is almost white, especially the lower jaw area. The oblong snout is clearly divided from the head. It is very narrow at the tip, which is curved slightly upward. Although very small, its eyes are functional and are not as degenerated as those of *Platanista minor*, the eyes of which have almost completely lost their function. The eyes are located on the sides of the upper part of the head. The large pectoral fins are characterized by a

roundish tip. The very short and triangular dorsal fin has front and back sides of equal length.

At one time it was thought that Chinese fishermen called this species the white-flag dolphin because when immersed in the water, its dorsal fin, breaking the surface, resembles a flag. In reality, this was not true and was due to confusion caused by the similarity between the words for *fin* and *flag*, and the name used today is white-fin dolphin. The slightly rectangular blowhole is longitudinally located on the left side of the crown. Each jaw has between 30 and 35 small, conical teeth.

At this point Western researchers have been able to examine very few of these species and knowledge of them is very fragmentary. All those studied from an anatomical point of view possessed a characteristic that can be found in other Odontoceti but is extremely striking in this one. The body is asymmetrical, the right side being much more developed than the left.

One of the difficulties in studying this species is its very limited geographical habitat. It can be found only in Tung Ting Lake, passages near the lake, a few of its tributaries, and in part of the Yangtze River. These are all located within China.

The ideal habitat for *Lipotes vexillifer* is shallow freshwater bodies with muddy bottoms. In fact, it is from the murky depths that this species draws its principal food. Observers have revealed that these dolphins sift through the mud in order to catch the fish that live on the sandy bottom. From the surface it is possible to see the water muddied by the movements of these animals. The fish found in the stomach of these species (when an autopsy was carried out) were those typical of muddy depths. Chinese fishermen have stated that sometimes they can hear the sound, similar to a bubbling, the dolphins make as they sift through the water and mud in search of food. This search for prey occurs most frequently during the hours of sunrise and sun-

set. In addition, *Lipotes vexillifer* seems to take advantage of its prey's geographical and seasonal distribution.

Large numbers tend to congregate in water passages where there is the greatest concentration of fish. Generally their schools are not very large. Schools sighted have ranged in size from 2 to 6, to maximums of 10.

Like other Plantanistidae, this species is a slow swimmer and can remain underwater only for short periods, usually a maximum of a few minutes. Then it rises to the surface to breathe. Its long snout is the first thing to emerge, followed by the head. Its blowhole, which is located on the crown of the head, supplies air to the lungs. After the head, its back emerges with its characteristic dorsal fin, which can be seen for a few seconds. Then the animal returns beneath the surface, usually without raising its tailfin above the water.

These dolphins do not seem to undertake true migrations, but probably move according to the need for food and changes in the water level. In spring, when the water level rises, the dolphins go upstream to tributaries and mating occurs. When the water begins to fall during the winter, they return to Tung Ting Lake and the Yangtze, where the females give birth between January and February.

Gestation lasts about 10 months, after which is born one offspring (this is true of all cetaceans). Especially in the first months, the mother remains very close to the newborn, when underwater or above the surface. If they become accidentally separated, due to the passing of a boat, for example, the mother becomes extremely watchful, continually turning her eyes

The rare Beiji dolphin has been completely protected by the Chinese authorities since 1975. The size of the population is unknown and is believed to be small (perhaps fewer than 500 individuals).

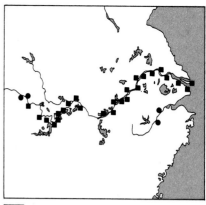

■ *Distribution area of the Beiji dolphin 30 to 40 years ago*
● *Distribution area in 1983*

toward the spot where her offspring was, refusing to stop the surveillance until the young is once again visible.

A specimen captured in 1980 was kept in captivity at Wuhan's Institute of Hydrobiology for 4 years. Researchers were able to make many observations regarding its behavior. As has been noted in nature, in this case also it was found that the specimen was most active at sunrise and sunset. The dolphin swam quickly, often changing direction, turning on its side, and flopping upside down. After meals, on the other hand, the specimen swam very slowly, regularly and unvaryingly diving under the water. Several times this dolphin displayed a behavior typical of many other cetaceans: it emerged vertically from the water, its entire head above water. Remaining in this position for a few seconds, it would turn its head, curiously surveying the surroundings. Daily this dolphin consumed a quantity of food equal to about 10 to 11 percent of its body weight, a maximum of 16 percent in the winter and of 7 to 8 percent in the summer. When being fed fish, this cetacean would always move itself in such a way that the fish entered its mouth head first, thus making it easier to swallow its meal.

A superstition held by Chinese fishermen once protected these animals from indiscriminate hunting: it was believed that anyone who caught one of these creatures voluntarily would come down with a serious illness. At the same time, dolphins caught accidentally were nevertheless taken advantage of in two ways. Part of the meat was eaten by the fishermen, and the oil was used to make medicinal concoctions. However, with progress, these superstitions have disappeared, and a certain number of dolphins are caught every year. Other, indirect factors have also threatened their existence. Often, for

The Beiji dolphin is a river dolphin with a long snout and an asymmetrical body, the right side being larger than the left, a characteristic found in other Odontoceti but extremely striking in this one.

example, these dolphins become entangled and die in fishing nets. The catching of fish in great numbers in the waters inhabited by this dolphin has resulted in a reduced food supply for this species. In addition, increased industrial development during recent years along the banks of Tung Ting Lake and its tributaries has compromised their survival. The water levels of the lake and river have been periodically controlled by humans, upsetting the natural habitat of this species, reducing its range, and disturbing its biological activities (reproduction, feeding, and so on).

Several dolphins have been killed by deep head wounds, probably caused by boat propellers. It seems that these dol-

phins were attracted by the noise made by the boats' motors and drew closer to the craft, receiving mortal blows. Only in the last years have Chinese researchers begun to study this species, and unfortunately it is still not known how many of these animals exist in their natural habitat. However, the population is not considered to be very large and numbers perhaps fewer than 500. Since 1975 the Chinese government has conferred complete protection upon this cetacean, but the damages caused by accidental captures and industrialization of the region have certainly not ceased. Because knowledge of these animals is very limited, more studies are absolutely necessary to improve the protective measures already adopted.

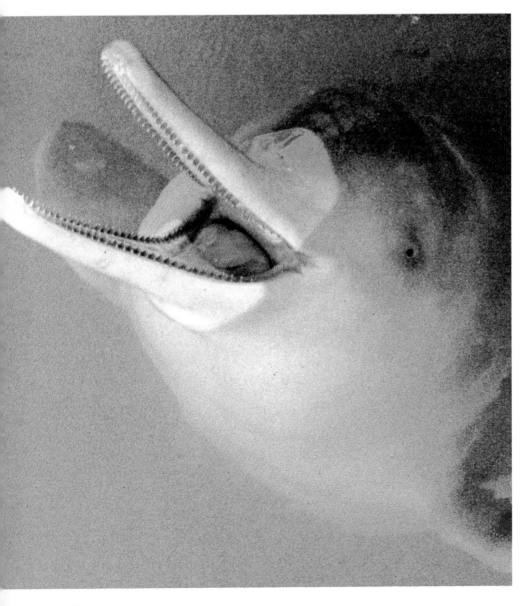

PAINTED TERRAPIN
(Callagur borneoensis)

Ex	R	E	T	**V**	I	K

DARK SOFT-SHELL TURTLE
(Trionyx nigricans)

Ex	**R**	E	T	V	I	K

PHILIPPINES CROCODILE
(Crocodylus mindorensis)

Ex	R	**E**	T	V	I	K

Found in the Indonesian archipelago from the Malay peninsula to the islands of Borneo and Sumatra, the painted terrapin *(Callagur borneoensis)* is a difficult turtle to find in fresh water because its habitat includes brackish tidal waters near coastlines. The carapace and limbs reflect its habitat; the carapace is very flat and the limbs are palmate. Its shell, about 40 cm (16 in) in length in the adult, is brightly colored and adorned with three large longitudinal black stripes. The background color is black with small yellow spots scattered on each plate. Smaller stripes spread out from each spot like rays. This motif is repeated, less regularly but distinctly, on the surface of the plastron. Specimens of this species have a rather small head with a serrated, horny beak for tearing food. The species eats plants most of all but also includes aquatic invertebrates in its diet.

The painted terrapin is locally protected by laws that prohibit its capture for commercial purposes, and the collecting of its eggs, a food source for the local population, is limited to certain quantities.

As with other Trionychidae species spread in southern Asia, the carapace of this aquatic turtle is decorated with light and dark concentric rings that are brightly colored, especially when young, and that disappear almost completely with gradual darkening in advanced years.

This is one of the species best adapted to aquatic life, and for this reason it has fully webbed feet. The carapace, very flat and soft because of the absence of horny plates, is about 70 cm (27 in) long. The neck is rather long, and the front of its head has a small snout with nostrils. This excellent swimmer and predator of the Bangladesh territory rivers, where it feeds on fish, mollusks, and other invertebrates, can be seen on land only at egg-laying time and during hours of thermoregulation. During winter hibernation, which lasts for several months, the turtle remains in the water.

The recent numerical drop in the numbers of *Trionyx nigricans* is caused by the massive affect of humans, who eat great quantities of these animals. In spite of the protection given the species, its exploitation in these areas puts it in serious danger of becoming extinct.

It is very difficult to distinguish this species of small crocodile from its close relative, the New Guinea crocodile (*C. novaeguineae*), and only knowledge of the place of origin truly helps in the classification of specimens. At the time of its discovery, which took place around the end of the 19th century near the Catuiran River on the island of Mindoro, it was immediately confused with the New Guinea crocodile, and the problem remained, despite the scientific differentiation made in 1923 by Schmidt, who had been able to study three young specimens of this species from Mindoro, south of Luzon. There are different opinions even today among various taxonomists so that it is possible to describe it at the same time as either a separate species, *C. mindorensis*, or as a subspecies of the other one, *C. novaeguineae mindorensis*.

There are in fact very few morphological visible external differences: the *mindorensis* has a wider and heavier cranium, with a less prominent triangular protuberance in front of the orbits. As usual, there are the small postoccipital shields, four on each side in transverse series, and the four large neck shields that form a square, on the sides of which are laid two very small shields. The color is also similar: olive-brown with almost blackish spots in the adults, paler colors with dark bands on the tail in the young.

The two species, or subspecies (according to the different points of view), are well differentiated as far as the area of distribution is concerned: for the *mindorensis* the fresh waters of the Philippine Islands of Luzon, Mindoro, Mindanao, Busuanga, and Jolo. Fossil findings show, at least in part, a common origin. Probably,

Distribution area of the painted terrapin

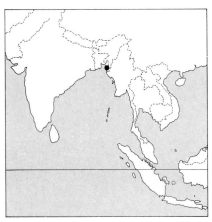

Distribution area of the dark soft-shell turtle

the original form colonized the various Pacific Islands; then genetic differentiation took place due to the distance between the populations (we must keep in mind that the two distribution areas are 2500 km [1550 mi] away from each other). Other theories, however, claim that the many common characteristics are a phenomenon of convergence in consequence of the adaptation to similar environments and the struggle to escape from the attacks of the estuarine crocodile (*C. porosus*).

Even though it would be impossible in nature, reproduction between the two forms has produced fertile individuals in captivity. Many doubts remain about its scientific classification, and as many about its biology, too. We know it prefers fresh waters, but it is also present in salt waters. The fertile, mountainous islands on which it is to be found have a tropical climate with a luxuriant vegetation that often makes the banks of the water courses impenetrable. Here, on the waters under the

No information is available on the populations of the Philippines crocodile. The species is suffering a strong decline in various parts of its distribution area.

"green galleries" the Philippines crocodile hunts its prey, above all fish, but saurians and amphibians are also a very frequent resource, often integrated with a few mammals that have incautiously ventured into its range: rodents, some mongooses, and at times even the tamarau (*Anoa mindorensis*), a small, short-horned buffalo of Mindoro.

Although its size is not exceptional for a crocodile—it is between 1.6 and 2 m (5 to 6 1/2 ft) in length—it is one of the largest animals in these areas, and thus when it reaches adulthood it is at the top of the food chain and has no natural enemies. The young may succumb to attacks from the countless occasional predatory animals and perhaps even to the parents themselves.

There is no data on the current size of *Crocodylus mindorensis* population, but a strong decline in numbers has been confirmed in many areas. It cannot be denied that annual climatic factors may massacre these reptiles. In that paradisiacal land, furious typhoons frequently devastate entire islands, causing floods and dragging along in the tumult of waters young and adult individuals toward the open sea. The major cause of its decline, however, is essentially human. If some populations are near extinction, it is because of the great demand for the skin, which leads unscrupulous hunters to kill on a large scale. The importation of young individuals for domestic breeding in European countries and North America is also considerable.

The Philippines crocodile has been put on the IUCN *Red List* and is listed in Appendix I of the CITES, but it does not yet have any official protection in the territory of the Philippines, and this absence of local laws makes its future less certain. As its biology is not yet known exactly, it is not even possible to start a serious program of intensive breeding so as to have biogenetic reserves on which to draw for future reintroduction in protected areas. Very few specimens are kept in captivity in zoos, and since the exact provenance of the specimens is not always known, there is no certainty that the specimens actually belong to this species.

SIAMESE CROCODILE

(Crocodylus siamensis)

Ex	R	E	T	V	I	K

Toward the end of 1700, during a stay in Siam (the current Thailand) some French missionaries observed particular crocodiles and sent a sample cranium to Europe. Only in 1801 was scientific validity attributed to the species, even if the Siamese crocodile *(Crocodylus siamensis)* was still confused with the mugger *(Crocodylus palustris)* for a long time. Actually, only young specimens are very similar, whereas the adults are substantially different, especially as far as their head is concerned. The muzzle is less broad in the Siamese species, and it has a middle ridge on the brow between the eyes. This loricate measures about 3 m (10 ft) in length, even though very old individuals may be more than 3.5 m (11 1/2 ft). Its favorite habitat is the edges of swamps and coastal lagoons, sometimes even with salt waters.

Although it is not normally found in sea water, the Siamese crocodile was able to colonize the little mountainous islands of the Pacific, where it lives even at remarkable heights. Its distribution includes Thailand, as far as upper Mekong; Kampuchea, in the northern part as far as Tonle Sap, and also reaching altitudes 1000 m (3280 ft) on the Luang Bian Plateau; Vietnam; and the Malay peninsula, south of the Patani River. The major islands where a few colonies of the species have been recorded are those of Java and Borneo. It is a matter of regions with a very sharp seasonal alternation. Long periods of drought are followed by periods of torrential rains and often real floods. This does not seem to jeopardize the annual normal activity of the *C. siamensis*, which has been observed in both seasons without periods of estivation. At the beginning of the humid season this reptile car–

About 200 specimens of the Siamese crocodile are thought to exist in the wild. This animal is also raised for commercial reasons.

Distribution area of the Siamese crocodile

ries out the laying of eggs.

Information about its biology is scarce, and the few studies have been carried out on the Thai populations. The offspring, which may be found in August, are a little under 25 cm (10 in) long, and their typical features are the very pale dorsal color scattered with dark stripes and spots, slightly wider than those in the young of *C. novaeguineae*. In Thailand, these crocodiles are commonly distinguished from the *C. porosus* by the degree of edibility: as the *C. siamensis* lives in fresh, flowing waters, it has a better "taste" than the mugger, which prefers muddy, salty waters.

The ceaseless hunting, especially ruthless, organized hunting for its skin, has been fatal to this loricate, which today perhaps survives with little more than 200 free individuals, mainly confined to the province of Nakorn Sawan (Thailand). No measure to protect them has yet been taken, except for inclusion in the CITES.

For this species, fortunately, the reproductive potential in captivity is high, and there are already farms for the breeding of *C. siamensis* and *C. porosus* near Bangkok (Thailand), although until now only for trade purposes. In the future, some of these animals might be useful for suitable reintroductions in nature.

ESTUARINE CROCODILE

(Crocodylus porosus)

Ex	R	E	T	V	I	K

As can be deduced by its name, this is one of the most aquatic among the crocodiles, certainly the greatest lover of salty water, in which it frequently swims, sometimes covering great distances. It may perhaps be compared with the ancestral Teleosaurian, and the partial reduction of the bones of the limbs may be seen as a first step toward a pelagic existence. The legs are in fact used only for helping determine direction and cause as little friction as possible.

Other somatic characteristics include its long muzzle, which is about twice as long as the width at the base, and two wrinkles that go in front of the eyes and converge on the nose. These animals are divided into two subspecies because the *Crocodylus porosus porosus*, existing in Ceylon only, has two rows of postoccipital scales that join together between the head and neck: these are not found in *Crocodylus porosus biporcatus*.

During the mating season, the males engage in ferocious fights, giving out roars and grunts that are not to be confused with those of the males of *Rana cancrivora*, whose distribution areas are the same. Then the female builds a nest, piling up plant detritus: sometimes this is decomposing material, but in general she pulls out grass and aquatic plants. As the distribution area is widespread, there are differences in the period of laying and in the distance of the nest from water, in order to avoid floods. The female watches over the eggs, dampening the area around them. They range in number from 25 to 72. From them hatch young of a gray-green color, black spotted on the body and tail. These young, having a longer muzzle, feed especially on small fish. Little by little the spots grow larger, and the adults become uniformly dark, even though some traces of the spots remain on the sides and on the tail. The growth takes place rapidly in the young, so rapidly that at three years of age they are already 1.5 m (5 ft) long, whereas it slows down afterward to about 25 cm (10 in) every two years. The total average length is 3.65 to 4.3 m (12 to 14 ft), but rare specimens have been known to grow to 6 m (20 ft). The adults prefer frogs, grass snakes, and small mammals; they rarely attack large animals, but they are the only crocodiles known to have carried out frequent predatory activity against humans (with documented attacks), to the point of attacking small boats.

These crocodiles are found above all in mangroves swamps and in estuaries of large rivers and live in an area that includes Sri Lanka, the coasts of India and Pakistan, the archipelago of the Sonda, the Philippines, the Moluccas, Palau (New Guinea), the New Hebrides, the Australian northern coast, and all the islands of these areas. Tireless swimmers, they have taken themselves hundreds of kilometers from the coast, and a specimen was found in Fiji. Not competitive with the other members of the family, but hunted for the enormous value of its skin, it is now in serious danger of extinction. In Sri Lanka, India, and Australia hunting of this species is illegal and exportation is forbidden. In Papua research is being done, and only special licenses are given. We are also trying to carry out an educational campaign, but national parks and effective laws are still necessary.

A tireless swimmer in the mangrove swamps and mouths of the rivers that make up its habitat, the estuarine crocodile can swim in the open sea as far as hundreds of kilometers from the coast.

without bending its head sideways, because this would cause problems in the water current.

Its favorite habitats are large rivers, including those subjected to periodic floods, such as the Barani, the Sadong, and the Inderapura. Its distribution area, which is rather wide, includes Borneo and Sarawak in Indonesia, Sumatra, and the south of the Malay peninsula. Apparently the species finds its optimal habitat in Indonesia, but it is very rare in Malaya and is extinct in Thailand.

Little is known of its way of life and nothing about the building of the nest. The eggs are large, measuring about 10.5 cm (4 in) in length and 7.5 cm (3 in) in diameter. From them hatch young of an olive color, with black crosswise stripes on the back and tail and black spots on the sides.

The greatest danger this species faces is not represented by hunting, but by the uncontrolled diminution of the natural environments due to deforestation. Included in the CITES, the *Tomistoma* is now protected in the states of Selangor, Negri Simibilan, and Malacca, and the national park of Taman Negara has been made for it. The Malay government also wants to introduce effective laws of protection and start breeding in captivity in western Malaysia.

Numerous specimens are being held in captivity in various zoos, above all in Bangkok (Thailand), where there are 170 specimens that will be possible to use for breeding.

GHARIAL

(Gavialis gangeticus)

Ex	R	E	T	V	I	K

This animal, the only representative of the family Gavialidae (whose first fossil remains go back to the Tertiary period) was scientifically described in 1789 by Gemelin, who gave this reptile the name *Lacerta gangetica*. It was later christened *Crocodilus gavial*, leading to the genus *Gavialis*. The name *gavial* comes from a mistake in writing the Indian word *gharyial*, a holy name that stresses that this crocodile was the son of the god Ganges. Since the European taxonomists had but little contact with these regions, the mistake remained and has become in many areas this animal's name.

As early as the 3rd century B.C., a Roman narrator describing the Ganges said it was home to a particular crocodile whose characteristic was a large growth on the tip of its nose. This description stuck and remained in the imagination of the classicists. It is true that large males sometimes have a small protuberance deriving from the inside of the premaxillary bone. Some people thought it was a growth to help them stay underwater longer, but apart from the fact that its size is limited, its presence only in old males makes such a theory extremely dubious.

The gharial, generally considered to be the most aquatic of the Crocodylia order, has special adaptations for this kind of life. Its average size varies from 3.6 to 4.5 m (12 to 15 ft) in length, but it may exceed 6 m (20 ft). The record was held by a specimen in captivity: 6.55 m (21 1/2 ft) long. The typical muzzle has a narrow, cylindrical nose, which is about four times as long as the base of the cranium. In the young it is even longer proportionally, being five times as long as the base of the cranium. The teeth,

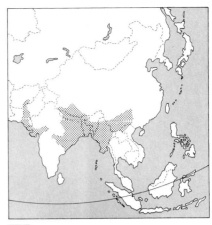

Distribution area of the gharial

of a greater number on the jaw than on the mandible, are the longest and sharpest among the crocodiles. The limbs are very weak in the adults. The hind legs, and in a lesser degree the front also, are built strictly for swimming and are webbed. After all, hardly ever do these animals abandon the water, except during the reproductive period. The long, thick tail has a crest so that it can easily serve as a rudder. Other features are the rather slender skeleton and the very thin scales, so thin as to be as sensitive as human skin.

In the distance, it seems of a uniform color, pale on the sand and blackish in water, but observed at close range it tends to become darker from the nose to the tail, even though it maintains an olive green eyes color on the back and yellowish on the stomach. The old lose their brightness, but the typical green eyes remain, ringed, with a narrow pupil split by a black vertical line.

Living always in water and being incapable of entering in estivation or hibernation, the gharial needs large rivers; moreover, with its particular shape and long head and thin neck, the rivers must not have very powerful currents. The optimal habitat, therefore, is the Ganges, from which the name *Gavialis gangeticus* derives. Its distribution area, however, is wider and includes different river systems: that of the Indus in western

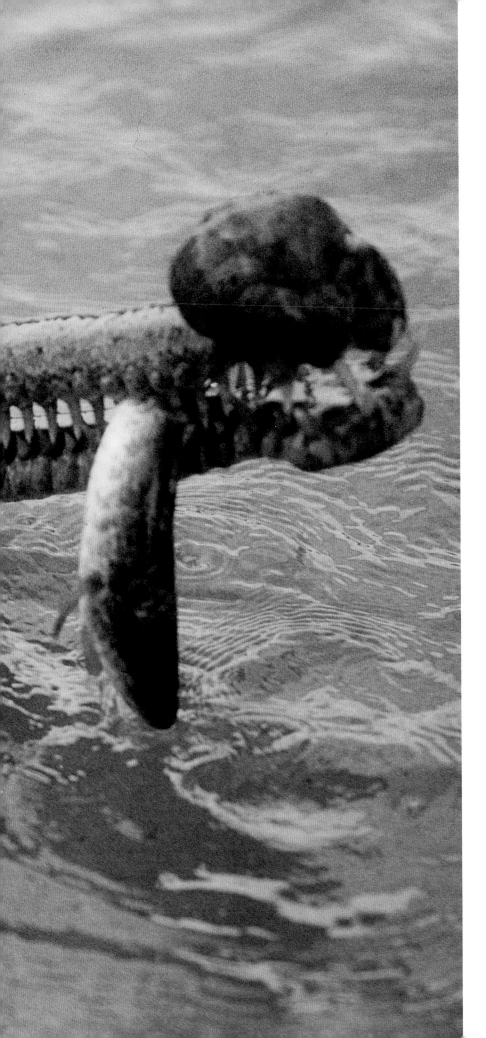

Pakistan, those of the Mohanadi and the Bhima in India, that of the Brahmaputra in eastern Pakistan, of the Assam in east India, of the Kaladan in west Burma, and that of the Irrawaddy in central Burma. Some individuals of the species sometimes reach Nepal and Bangladesh through the Ganges basin. Because of its slowness, the gharial has never been able to reach the systems of Godavari, Tapti, and Nerbudura, even though their northern tributaries originate very near the sources of rivers that flow into the Ganges. Their distribution toward the western lands of the Hindu has been stopped by mountains and deserts.

Because of its indolence and laziness and, above all, because of the fact that, even during the driest season, the large rivers never dry up, we may reject the supposition made by some naturalists, according to which the gharial migrates. These river waters are rich in its favorite food, which is practically its only food: fish. At times, large adults may also capture aquatic birds and small mammals. Its technique of capture, which has been observed at close range in young individuals kept in captivity, is unique. Some of these were put into a pond with a current. Fish of the family Cyprinidae were introduced, and they swam together in a school. One of the gharials chased them until they were cornered at one end of the pond. Here it stopped for a moment, as if waiting, and then, with a movement too quick for the human eye, snapped at them. Paying great attention, it is possible to see that only the head and neck are used, whereas the rest of the body remains still; in

The protuberance on the end of the nose of the gharial, known from antiquity, is found only occasionally in very large males. This is the most aquatic of the crocodiles, and it eats almost exclusively fish. The illustration shows a specimen with captured prey.

75

movement the head almost forms a right angle with the body, so as to grasp the prey with the side of the fauces. With its bite it usually catches one fish at a time, rarely two, rarely none. Other observations were carried out on an individual that was also in captivity, but alone. Using again a school of fish as prey, it was observed that the gharial snapped only when they passed near its right side, whereas they were ignored on the left. It was surprising to notice that such an ancient animal, often considered intellectually weak, has acquired one of the most typical features of superior animals, and above all of man, that is, the greater development of one of the two encephalic halves.

In the stomachs of some examined specimens numerous metal objects and human decorations have been found. For this reason it was labeled a maneater. Later, it was possible to ascertain that this reptile is harmless to human beings. Recently, the hypothesis has been made that the gharial swallowed these objects when digging in the sand and in rubbish heaps. It is necessary to consider that into the Ganges, the holy river, are thrown decorations in sign of worship, and that on its banks are set up pyres where corpses are burnt and where are laid all the jewels of the defunct. According to another hypothesis, the objects may be swallowed voluntarily with a gastric function. Real proofs have never been obtained, and this may all be a matter of legends that have been mixed into research actually made on the Nile crocodile.

Little is known about the natural habits of these reptiles in their environment. After mating, which takes place in the water, the female is said to dig a hole in the sand of the river banks that has the function of a nest. In it she is said to lay 40 eggs in two groups of 20 eggs each, separated by about 30 cm (12 in) of sand. Even this information may be little more than hearsay, though, because in more recent studies the female has been seen to build a nest in which she also

laid plant remains, as do other crocodiles, and to leave in it from 20 to 95 eggs, at a depth of about 50 to 60 cm (19 to 23 in) and about 1 to 5 m (3 to 16 ft) away from the level of water, always during nocturnal hours. The laying generally takes place in March and April, but more in-depth research is needed on this subject, for, in the greater part of India, rains are brought by the southeastern monsoons. These winds first reach the coasts and then the internal region during the period between May and October, so the young would be born in the driest period of the year. The young gharials live on the beaches, in stagnant waters, in small isolated puddles or irrigation areas, leaving the larger rivers to the adults. This might depend on the enormous difference in size, rather than on possible cannibalism, which has never been noticed in this species. The occupation of a different ecological niche might be therefore a response to the necessity for different kinds of prey.

The population is today in rapid decline. It seems that in India there are about 100 individuals only, in Pakistan there are probably not even 10, in Burma and Bangladesh they might be already extinct, and a few other individuals are probably to be found in the few areas left. The gharial has no natural enemies. Competition with the estuarine crocodile is largely precluded by its specialized diet, and that with the mugger is impossible because besides diet differences there are habitat differences. This means that man alone is responsible for its decrease. Besides the hunting to obtain their skin, many individuals are accidentally killed by the dynamite used for poaching.

A gharial on the banks of a river. The idea that the females deposit eggs in two holes in the sand of river banks may be only hearsay. Recent observations include the building of nests, as with other crocodiles, including the use of plant parts.

The gharial is included in the CITES, and since 1972 India, Bangladesh, and Pakistan have officially protected it, forbidding its hunting and the exportation of both live individuals and their skin. Moreover, 1500 young animals were gathered in India and put in protected areas in Pradesh, Orissa, and Rajasthan. Plans for the future are being considered, especially in India, where, under the control of expert herpetologists, there is the intention of gathering eggs and young individuals to breed them in semi-captivity and reintroduce them in protected areas.

Numerous specimens are in zoos, where they live long lives; in 1937 a specimen was still alive after 24 years in captivity.

A great problem is that, above all in America and Europe, the reproductive potential of the gharials in captivity is very low.

CHINESE ALLIGATOR

(Alligator sinensis)

Ex	R	E	T	V	I	K

Distribution area of the Chinese alligator

The first European to speak about this alligator, which is called *Yow-lung* or *Tun-lung* in China—both names means "dragon"—was Marco Polo. It was scientifically described by Fauval at the end of the 19th century. Its size ranges from 1.2 to 1.8 m (4 to 6 ft), but individuals in captivity have reached almost 4 m (13 ft). The muzzle is proportionately sized, with a shorter nose than the very similar American alligator. The squat body ends with a short tail, and the weak limbs and non-webbed feet make their movements on land even more awkward than those of crocodiles. There are plates above the eye sockets, similar to those of the caimans. Other alligators do not have these plates. The back is a black-green color, and there are sometimes white and gray stripes on the stomach. The young differ from the adults by being brightly speckled with yellow and black, and they also have five yellow crosswise bands on their bodies and eight on their tails. This reptile's very powerful jaws are an indication of diet. It feeds preferably on aquatic turtles, which are, in fact, abundant in the waters in which these alligators live.

With the advance of human settlements, this alligator has turned into a nocturnal, predatory animal of chickens and small dogs. Little room is left for it in the valley of the lower Yangtze, where the large river annually floods its banks and turns the maze of channels and pools into grassy water meadows and swamps, uninhabitable for humans and rich in natural shelters. Here, it can easily dig dens in which to hibernate during the winter, and the females can build nests, which they construct on small islets a few meters from the water, so as not to let them be flooded, where arboreal vegetation offers shade.

With sideways movements of the body and tail, the female heaps up dead leaves, mud, grass, and other nearby debris. Then the female uses the hind limbs, working alternately, to dig a hollow on top where she lays from 20 to 40 eggs. She also crawls atop the nest to close the opening, paying close attention not to crack the eggs. The complete nest measures about 1 m (39 in) at the base. The female remains for the entire time on guard at a little distance away or even with her throat resting on the construction. She goes into the water often and then come back out and shakes water off her body to wet the ground and maintain the correct temperature. After hatching, the young reach the water on their own, attracted by instinct to the brighter or wetter area.

In Siam (now Thailand), these animals were once caught and bred for the sovereigns' banquets, and there is evidence of alligators having been exhibited in fairs in Shanghai. It is said the Buddhist priests, who used to breed these alligators in their temples, gained merit by freeing them. Today, sought after for their skin and

With its short nose and powerful jaws, the Chinese alligator's favorite food was aquatic turtles; with the growth of human interference, it has taken to preying on chickens and other small domestic animals.

meat, these alligators are virtually extinct.

Some very small populations have probably survived in the provinces of Anhui, Chekiang, and Wuhu, where there was once the greatest concentration of specimens. No conservation measures have been proposed apart from inclusion in the CITES.

Sought for its hide and its meat, the Chinese alligator is nearly extinct: a minimal population may yet survive in the wild.

SAIL-FIN LIZARD

(Hydrosaurus pustulatus)

Ex	R	E	T	V	I	K

As indicated by its name, the members of the *Hydrosaurus* ("water lizard") genus are adapted to an aquatic environment. Their most characteristic trait is a crest of skin, which is supported by long bony projections from the caudal vertebrae starting at the base of the tail. This crest, which resembles a sail, can be as tall as 6.5 cm (2 1/2 in). With its undulating movement it functions as an auxiliary paddle for swimming. There is another, much smaller, crest on the lizard's back; this is a characteristic peculiar to the Agamidae family, to which this genus belongs. The secondary crest, made up entirely of scales, runs from the base of the tail up to the neck. As is frequently the case among the Sauria, there is marked sexual dimorphism, in that the females have no crests at all, and their coloring is less showy than the brilliant green-brown of the males.

The rest of this lizard's body is also adapted to its environment. It is muscular and has a flattened trunk and tail, rendering it more hydrodynamic. It has a characteristic fringe of scales on the toes of both the front and hind legs, making the toes lobed. These membranes are particularly large on the hind legs, allowing the animal to run for short distances on the surface of the water in a semierect position, holding itself up on its hind legs, when fleeing predators. Some of the Iguanidae of the genus *Basiliscus* of South America also display this interesting characteristic. They, too, move on two feet when in flight, using their long tail as a balance and moving their hind legs so rapidly that they manage to walk on water for brief stretches. This is clearly an example of the phenomenon of adaptive covergent evolution due to the similar environments inhabited by the two animals. In any case, it is easy to trace their descent from the Ornithi-schian (bird-hipped) dinosaurs. The *Hydrosaurus* is also known as the water dragon.

There are other similarities and differences between *Hydrosaurus* and *Basiliscus*. It can be noted that they are similar in coloration and body form, each having a developed crest. Anatomically, however, the basic difference lies in the fact that while the *Basiliscus's* teeth are set in the inner side of its jaw, and it has replacement teeth, the *Hydrosaurus* has its teeth set on a bony maxillary crest, and they are practically final, that is, without the possibility of replacement.

The *Hydrosaurus* lives in primary and second-growth forests along the rivers and freshwater waterways distributed among the Philippine Islands (Luzon, Polillo, Mindoro, Negros, Dinagat, and Mindanao) and the other islands in the surrounding sea: Cebu, Bohol, and Panay. This archipelago of volcanic origin has some of the richest and most luxuriant environments on earth; palm trees and unique plants of every kind border the waterways and pools that fill the craters of the extinct volcanoes. These unusual little reptiles move through the vegetation in search of their favorite fruits and leaves, often catching centipedes and insects, which are abundant in this region. Almost nothing is known of this reptile's behavior in the wild, in part because of the difficulty in extracting them from or penetrating into their environment and in part because of the "semi-aquatic" habits and their ability to camouflage themselves.

Mating probably occurs, as in other tree-dwelling Agamidae, among the foliage, where it is the female who expresses territoriality and communicates her availability to the male by raising herself up on her hind legs and arching her back. The male grips her by the neck with his mouth and brings together their respective cloacal openings for copulation, which lasts about one minute, with both

A slow-moving animal unafraid of humans, the sail-fin lizard is threatened by hunting for its meat. It lives in forests along rivers on various islands of the Philippines

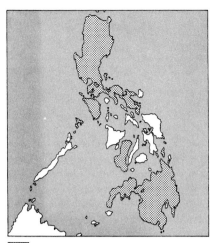

Distribution area of the sail-fin lizard

INDIAN PYTHON

(Python molurus)

Ex	R	E	T	V	I	K

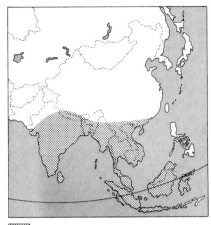

Distribution area of the Indian python

animals completely immobile. The female lays four to six eggs in a little hole dug in the vegetation that she then covers with turf, using her head and her hind legs. The young reach maturity approximately a year and a half after birth, but while the females remain in the area where they were born, the males go in search of territory free from competitors of the same sex. The first few years of life are the most perilous for these lizards, and during this period the young are remarkably retiring and apt to flee. They remain hidden as much as possible to avoid attack from predators, including rats and water snakes. Even after they are fully grown—they can be as large as 90 cm (35 in)—they have many possible predators: mongooses and birds of prey pose the greatest danger, along with the *Crocodylus mindorensis*. As they are somewhat slow moving on dry land and have little fear of humans, the *Hydrosaurus* has always been hunted by natives, whose diet is based on this animal's firm, white meat.

It is unlikely that anything can rescue them from disappearing on the most developed and populated islands, but even though the proper steps for protecting this animal have not been taken, it may be possible to maintain a sufficient number of natural populations of these Sauria in appropriate reserves on the islands of Cebu, Bohol, Panay, and Negros.

Although the fantastic lengths with which it is credited probably do not correspond to reality—it is said to reach up to 7.5 m (25 ft) in length—this is without doubt one of the giants of the genus and, at least in Asia, it is second largest of all the snakes, after the reticulated python *(Python reticulatus)*, in terms of body size.

The species can be identified by its coloring and by the number of scales in the longitudinal dorsal series, and furthermore it has the hollows with thermoreceptive properties only in the first two upper labials on each side.

The species is subdivided into three forms, which are fairly well differentiated and can be found in distinct territories. The typical subspecies, *Python molurus molurus*, is widespread in Pakistan, India, Nepal, and Bangladesh; the subspecies *P. m. bivittatus* is found in Burma, southern China, and south as far as Malaysia, Indochina, Borneo, and Java; and *P. m. pimbura*, lives in Ceylon (Sri Lanka). Of the three, the Burmese python is the largest, with a thicker body and brighter coloring. The basic color is yellowish-brown or grayish-brown, with large subquadrangular spots of a reddish-brown color bordered by cream or gold. On the head an upside-down V-shaped mark of cream color can be seen. The race that gives the species its name has less intense colors and is of smaller size. For a long time these individuals were considered the dark phase of the species, whereas the Burmese variety was considered the light phase. The third race is the smallest, less than 0.5 m (20 in) in length, and also the hardest to observe.

Capable of adapting to a great variety of surroundings, the Indian python frequents the margins of pools and swamps, the banks of rivers, and humid forests, but also the grasslands between woods and cultivated land. Here they pass the daylight hours, above all the hottest hours, coiled up in a tangle of bushes or on the lower branches of trees. They become fully active only at dusk and at night, when they start to wander around in search of suitable prey. Assisted by a Jacobson's organ (with a strong sense of smell) situated on the roof of the oral cavity and capable of perceiving the thermal emanations of small mammals, they head without difficulty toward possible prey. These consist mainly of mammals, birds, and proportionately sized lizards, and often the domestic animals (goats, pigs, hens) that live near human dwellings. Pythons attack with lightning speed, gripping the victim and, if this is of a certain size, immediately wrapping it in coils of its body and, by means of a series of progressive tightening movements, suffocating it. After swallowing, if the prey was large, the python becomes torpid, heavy, and deformed. It therefore remains indolent and inert until

Right, the Indian python is the second largest of the Asian snakes.

The Indian python, second largest of all the Asian snakes, passes the daylight hours coiled in a tangle of bushes or on the lower branches of trees.

Also known as the desert or sand monitor, this is probably the bravest of the Sauria. In fact the inhabitants of the regions in which it can be found fear it, because, if surprised in the open with no means of rapidly hiding in its lair, it will first give warning to the disturbers by means of well-defined movements and will then attack men, horses, or camels, launching itself against them and raising itself up to 1 m (39 in) from the ground using its heavy tail. It is precisely this tail that distinguishes the species from others, as it has a rounded end, and is not streamlined. Other differences can be found in the oblique nostrils similar to cracks, and in the form of the incisors, which are small and

digestion is completed.

The reproductive period of the Indian python occurs annually and varies from colony to colony. Courtship rituals in the wild are not known, but the parental care later taken by the females has been noted. After finding a suitable place, shady and humid, the female lays the eggs in a single mass (between 30 and 100, according to size) and then immediately coils around them. Thus, in a type of incubation, the female remains throughout the period of development of the eggs, without eating or drinking. It has been ascertained that contraction of the female's muscles manages to raise the body temperature by a few degrees to maintain the incubated eggs at a temperature slightly higher than that of the surroundings.

Although they are protected by various national and international laws (Wildlife Act in India; Appendix I, CITES), the Indian python is seriously threatened today, both due to the reduction of natural habitat in which it can live undisturbed and with sufficient prey, and also due to the widespread commerce to which it is subject for its skin and for breeding in captivity.

wide. Typical of the family, on the other hand, are the head, which is longer than that of other Sauria; the tongue, hidden in a protective membrane when retracted and with two horny points; the absence of palatine teeth; and the absence of femoral and anal pores.

As suggested by its Latin name *(griseus)* the coloring is certainly not vivid. On a background of yellowish-gray, it has crosswise brown stripes on back and tail and also along the sides of the nape. The belly, on the other hand, is of a lighter color, tending toward yellow or beige.

This monitor was widespread throughout northern Africa and Southeast Asia, more precisely in the former Soviet Union (Turkmenistan, Uzbekistan, Tadzkikistan, Kirghizistan, Kazakhstan), Iran, Pakistan, Afghanistan, Palestine, and Arabia. Its favorite habitat is desertlike areas, abounding in rocks, which are sometimes crossed by small plains and sandy dunes.

In markets snake charmers sell

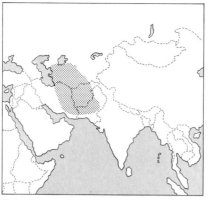

Distribution area of the central Asian monitor

these lizards to tourists, attributing to them extraordinary properties, after having treated them very harshly. With their teeth filed down, imprisoned in leather bags without food or water, to the misled buyer they appear a docile pet. However, it is true that, if reared from an early age, they can be domesticated and will even accept food from the hand, although the force of their jaws should never be forgotten.

For these reasons, along with

the progressive decrease in desert and isolated areas, today only two or three at the most of these rather timid animals can be found within a radius of several kilometers. The free individuals are protected by law in the former Soviet Union and in the nature reserves of Repetek, Bandkhys, and Tigrovaya, and are included on the CITES lists. The creation of other reserves in the uncultivated areas where the larger colonies are still to be found is also desired.

The aggressive central Asian monitor (below) lives in desertlike habitats or the steppes, always in rocky areas near water.

CENTRAL ASIAN COBRA
(Naja oxiana)

Ex	R	E	T	V	I	K
		E				

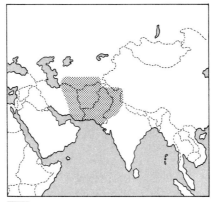

▦ *Distribution area of the central Asian cobra*

FIJI SNAKE
(Ogmodon vitianus)

Ex	R	E	T	V	I	K
				V		

The cobra is very common and widespread throughout intensely inhabited regions, such as Africa and the southern Asiatic countries. Tradition makes them the most poisonous and dangerous snakes in the world. This is particularly true, but only when taking into consideration the cases of death from bites. In reality, these ophidians are fairly indolent and not particularly aggressive, as the "snake charmers" well know. *Naja oxiana* is the form of cobra (very similar to the subspecies *Naja naja*) that is found farthest west, being present in the southern part of the former Soviet Union, in Afghanistan, and in Kashmir.

Unlike the case with other, similar species, the well-known hood (produced by the flattening and widening of the neck, as a consequence of the ribs in that area being elongated and mobile) does not have any dorsal markings; the back is of a uniform shiny black color, while the belly is yellowish. The size is notable and can reach over 1.8 m (6 ft). It preys on rats, mice, birds, and Sauria, which are killed in a few moments by the poison injected during its lightning-swift bite. Oviparous, it lays up to 22 eggs per year in a nest.

Although *Naja oxiana* has a wide distribution today, it is greatly reduced in numbers, following the destruction of its habitat and direct killing. Two reserves already exist in the former Soviet Union in which the species is protected, and its capture is controlled in the rest of the territories in which it can be found.

Along with such well-known and feared snakes as the cobra, bungarum, and mamba, the family of Elaphidae also contains numerous small but dangerous snakes. Among these is the American coral snake, fascinating for its varicolored markings.

These Elaphidae live a very secluded life, in the tunnels of tiny mammals and under stones and rotting tree trunks, and for this reason they are not feared as much as their larger Australian and Asiatic cousins. Yet their poison (although it can only with difficulty be injected into humans when biting, due to the small size of their teeth) is among the most powerful and is almost always mortal in effect.

The Fiji snake also has these characteristics: it is small, with a head not clearly distinguishable from the neck; it has small teeth, of which the two anteriors (on the jawbone) are connected to poison glands. The effects of the bite are neurotoxic and leave the victim with no chance. The external aspect of the snake is not as attractive as that of the *Micrurus* or the *S. coralli*. The surface of the back is dark brown or almost black. The lateral surface is of a lighter shade. In the younger animals, the head has yellow spotting, and the sides are brown or white with black spots. Among the morphological characteristics note should be taken of the extremely small eyes, with round pupils; the mouth shaped like an elongated dot, with upper and lower jaw aligned; the nostrils open between the first labial scale, the two nasal scales, and the internasal. The preocular scales are thin, and the postocular is single. The labial scales number six. The body is cylindrical with smooth scales, aligned on the back in 17 rows. The scales on the belly are rounded at

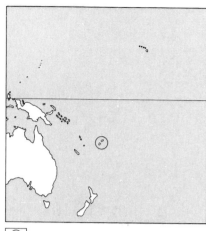

Distribution area of the Fiji snake

the edge and number from 139 to 152. The anal is double; the tail short.

Apart from the specimens in zoos, which, according to a survey carried out in 1971 and repeated in 1977, total 18, no estimate of the population in nature is known. In the Fiji Islands the preferred habitat is that of very thick forests, and the Fiji snake's presence has been ascertained on the island of Viti Levu, and probably also on Vanua Levu, Tave Uni, Kandavu, and Lau. It is not easy to observe them in these areas, because the primary forests are thick with hiding places, but it appears that the species is rapidly diminishing due to killing by local inhabitants, carried out particularly to limit the possibility of bites from these dangerous snakes, but mainly for religious motives.

Below, the Fiji snake

LATIFI'S VIPER

(Vipera latifi)

Ex	R	E	T	V	I	K

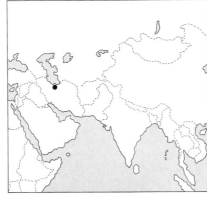

Discovered in 1966 by Dr. M. Latifi, this viper is for the moment known only in the Upper Lar Valley, in the Elburz Mountains in northern Iran, about 60 km (37 mi) northwest of Teheran. With a length of at most 78 cm (30 in) in the male and 70 cm (27 in) in the female, this viper is characterized by great polymorphism in its dorsal markings and by numerous morphological characteristics. The basic color of the back varies from brownish to gray-brown to yellowish-gray. It may have dark zigzag bands, diffuse dark marks, a thin vertebral line, or may even be completely without markings. The head has two large superocular scales that are separated from the eyes, with the classical elliptical pupil and tiny circumocular scales. This viper was initially considered a subspecies of *V. raddei*, from which, however, it is distinguished by the different number of ventral scales.

The areas inhabited by the species are divided between regions of typically steppe-like habitat, others that are rocky, and others of an alpine type with permanent glaciers; the viper is most common in the first type.

Courtship and reproduction have been observed halfway through June, and the females give birth around 3 months later to 5 to 10 young with an average length of 17.7 cm (7 in). In captivity, however, earlier reproductive periods have been noted, and births have taken place as early as July. Experiments on the heredity of dorsal markings, carried out by mating individuals with different markings and checking the predominance of one or the other in the young, have proved interesting.

Their diet is based on small mammals. The species is considered in danger and is on the

Distribution area of Latifi's viper

IUCN *Red List*, both because of its restricted distribution and because of the decision to use part of the Iranian valley as an artificial reservoir, with the construction of a dam and the consequent submersion of a great part of the area. Research is being carried out by the Wildlife Protection Society in Iran to evaluate the direct and indirect damage that will be done to the species by this serious alteration to the environment.

Furthermore, since 1883 mongooses have been introduced on all the islands, and these ravenous Viverridae have certainly had a serious influence on the numbers of the *Ogmodon*. It is not possible, therefore, to intervene with general protective measures without first eliminating this specific cause of reduction.

ISRAEL PAINTED FROG

(Discoglossus nigriventer)

Ex	R	E	T	V	I	K

The Israel painted frog (*Discoglossus nigriventer*) is an amphibian whose rather flat head is as long as it is wide. It has no parotid glands, and has eardrum membranes that are difficult to make out. The toes of the front limbs are free, but those of the hind limbs are webbed at the base. The total length of the adults is about 4 cm (1 1/2 in). The dorsal color ranges from ocher or rust to dark olive-gray; the ventral color is grayish-black with numerous white spots.

The species is aquatic, frequenting freshwater swamps, which it probably leaves only on particularly damp, rainy days. Little is known of the biology of *D. nigriventer* because only a meager amount of information has been gathered from direct observation. The eggs have never been observed but are believed to be dropped to the bottom of ponds in a way similar to that of other species of *Discoglossus*. The tadpoles are aquatic, and the point of their tail is bluntly rounded.

This species is seriously threatened, and, indeed, there is reason to believe that it may already have

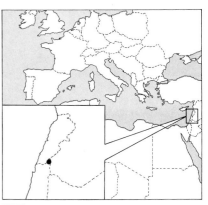

The possible distribution area of the Israel painted frog

become extinct in the wild. In fact, only very few specimens from Lake Huleh, in northern Israel, have been seen. The drying up of this stretch of water seems to have led to the disappearance of the species, although it is possible that some populations are in Syria in the areas bordering Israel. The causes of its decline are not completely clear, although the draining of the marshes has evidently contributed to the reduction of the populations.

D. nigriventer has been completely protected in Israel since 1940, the year in which the species was discovered, but no protection methods can be proposed. Aside from the fact that nothing is known of where the species can be found, nothing is known of the possible existence of specimens in captivity that could be used to reintroduce the species into its natural environment.

Birds

There are fewer nerve cells in the retina of the human eye than in the eye of a bird. In fact, sight is a bird's most developed sense; its eyeballs may weigh more than its brain. The importance of sight in the life of this class goes without saying: the falcon, which flies at dizzying heights and dives on the small prey it spots on the ground from far away, is a proverbial example.

The eyes of birds are usually positioned on either side of the head. Because of this arrangement, it is rather obvious that the field of vision is fairly wide, but, on the other hand, binocular vision, which allows exact perception of distance, is reduced. Composing a considerable exception to the nocturnal birds of prey are Strigiformes. In this order the large and immobile eyes are positioned on the frontal plane and are surrounded by wide cavities called facial disks. The crystalline lens is very developed, and the cornea is particularly convex. Moreover, a unique characteristic among birds, an opaque nictitating membrane is also present. The position of the eyes and the structure of the cornea gives the nocturnal birds of prey a field of vision equal to about 180 degrees, the central third of which is binocular vision. A bird's eyes are only slightly mobile, but this is usually compensated for by a remarkable mobility of the neck; the Strigiformes can rotate the head about 270 degrees (three quarters of a complete circle) without moving their body.

Endangered Asian birds involve more than 60 forms, and some Strigiformes are included among them. The name of the order, from the Latin *strix*, signifies "shape of the little owl." The owl, barn owl, tawny owl, scops owl, in addition to the little owl, make up part of this order of nocturnal predators, the general appearance of which is well known. We have already spoken about sight, but these birds also have an extremely refined sense of hearing. In complete darkness, even when their large eyes cannot see a thing, the nocturnal birds of prey, thanks to their sense of hearing, are able to sense the presence of potential prey, locate, and capture it. Moreover, they are able to pounce on small mammals completely unnoticed.

The various species are of assorted dimensions: the wing span, for example, varies from .3 m to 2.74 m (1 to 9 ft). Their soft and abundant plumage gives them an appearance of being quite large: a "plucked" tawny owl, for example, is equal to only about one eighth of its apparent size.

Almost one third of the endangered forms of Asian birdlife are Galliformes. The order appears to be rather large, if we recall some of the well-known forms: partridge, quail (the only species with migratory habits), wood grouse, peafowl, guinea hen, turkey, pheasants, and so on. The list of these few names demonstrates one of the reasons for the notoriety of some species: the beauty of their plumage. A good example is the splendid tail feathers of the peacock, which unfolds them like a fan and moves in parade in front of the females of his territory during the reproduction period. The peacock, along with many other Galliforme species, demonstrates sexual dimorphism; the difference in appearance between the male and female is remarkable. Most of the species have terrestrial habits and nest among bushes, their diet consisting primarily of vegetables. Their distribution is extremely wide and includes very diverse environments; there are Galliformes among in the icy Arctic tundra, as well as in the thick vegetation of tropical forests.

The order of Gruiformes is considered a neighbor to the order of Galliformes: it includes cranes, a rail, and an Otitidae, the great Indian bustard. The order includes several families and about 200 species whose forms are heterogenous. Those equipped with long legs, like the cranes, are also called wading birds, a traditional name that unites them to birds of the Ciconiiforme order. In general the Gruiformes always have rather long legs suitable for running over the ground, and short rounded wings with slight adaptation to flight (the crane is an exception). Making up part of the order, along with the others, are the great bustards and numerous groups of medium-sized birds with aquatic or terrestrial habits that are perfectly adapted to marshy vegetation or the banks of rivers and lakes. Among these are the rails, the most numerous families of which are the water rail, spotted crane, water hen, and coot.

The beautiful cranes, which lend their name (Grus) to the order, are slow and powerful flyers that fly in flocks in the characteristic V-formation. Under favorable conditions a flight speed of around 88 kmh (55 mph) has been recorded, and radar has identified migrating flocks flying above airplanes, at a height of more than 4875 m (16,000 ft). Pliny the Elder noticed their migratory habits: "they agree on the day of departure, they fly high in order to see far away, they choose a chief to lead them and, to the rear, they place guards that alternate in their monitoring, who keep the group intact by screeching." The knowledgeable Roman naturalist noticed that they sleep with their heads hidden below the wing while balancing on one leg.

SHORT-TAILED ALBATROSS

(Diomedea albatrus)

Ex	R	E	T	V	I	K

One of the few species of northern hemisphere albatross, the short-tailed albatross is seriously threatened by extinction and presently reduced to a few hundred individuals. This albatross is among those of medium size: 94 cm (37 in) of total length and around 213 cm (84 in) of wing span. The adult's color is white, except for the wing tips, which are black, and the tail, which has a black band. The bill, large in proportion to those of other species of albatross, is a vivid red, as are the legs, while the plumage of the head is a yellowish color.

The young are completely black when they leave the nest and become white, little by little, as they age. When they reach sexual maturity and begin to nest, they begin again to reveal a noticeable stretch of dark plumage, and within a nesting colony it is possible at a glance to determine the various ages of the brooding adults by the relative amounts of the black and white parts of their plumage. As with all the other species of its genus, the short-tailed albatross nests in crowded colonies on small ocean islands.

The mating system is strictly monogamous, with ties that are established upon attaining sexual maturity (at around 5 years) that last a lifetime. Even so, the couple spends only a brief period together at the beginning of the reproductive season, at which time they repeat the courtship ceremony and then mate the rest of the year, each of the two adults moves about on its own on the ocean. The system of cooperation between these birds during the brooding and raising of the young involves long periods in the nest and long, solitary fishing outings at sea, with very brief encounters, purely casual, at the nest, to which both adults return at intervals of several days to feed their young. The brooding of the sole egg and raising of the albatross chick take, in all, around 1 year and constitute therefore an extraordinarily heavy responsibility, practically without equal in the entire class of birds. The nesting of the short-tailed albatross, as indeed that of other albatrosses, takes place at 2-year intervals.

In the past, the short-tailed albatross nested on the islands of Torishima, Izu, Kita-no-shima, Nishi-no-shima, on all the Bonin islands, and on Kobisho, in the Senkaku group. They are also believed to have nested on the Pescadores, between Taiwan and continental China, on Agincourt, north of Taiwan, and at Iwo Jima, in the Vulcan islands. Presently, some 150 couples remain on Tsubakuro Point, on the southeastern part of the small island of Torishima, spread over barely 5 square km (35 square mi), and, in addition, some 25 couples are on the Senkaku group. Outside of the reproductive period, the short-tailed albatross, like all other representatives of its family, frequents the open ocean and is even able to travel far away from the small islands on which it nests, pushing itself as far as 10 degrees north latitude.

Until the beginning of the 20th century, the short-tailed albatross was still fairly abundant, especially on Torishima, where thousands were killed by feather collectors between 1887 and 1903. This collecting was officially prohibited by a government decree in 1906, but in practice the decree was never respected. A census carried out in 1929 on the island of Torishima revealed that the albatrosses there had been reduced to 1400, a number that fell to 30 to 50 10 years

Short-tailed albatross in flight (left): like all albatrosses, it can glide on wind currents for a long time without beating its wings. It nests in large colonies on ocean islands. Right, adults have white plumage, the young are black and become white as they age.

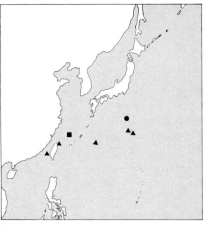

- **•** *Contemporary nesting sites*
- **▲** *Areas of sightings*
- **■** *Former nesting sites*

taken place on Torishima, in 1902 and 1939. Should a similar event occur within a short time, it would have a damaging effect on the local albatross population, which is still very small and very vulnerable.

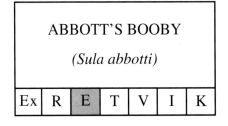

ABBOTT'S BOOBY

(Sula abbotti)

Ex	R	E	T	V	I	K

rather low reproduction rate, as do the other boobies. Each female lays only one egg every 2 years and does not begin to nest before 5 to 6 years of age. Under normal circumstances, this fact would not constitute a problem, for the mortality rate for the Abbott's booby is very much lower than for other species that have higher reproductive potentials. A global estimate of the population taken in 1939 to 1940 gave a total of 500 to 700 pairs.

later. Only at this point was feather harvesting interrupted, given that it evidently was no longer of economic interest. During World War II, the species was observed only once on Torishima, in 1945, and in 1949 it was prematurely believed to be extinct. In 1950, birds were again found nesting on Torishima, and their number subsequently increased slowly but regularly. Twenty-five specimens with 6 breeding pairs were counted in 1954, 12 breeding pairs in 1956, 26 in 1964, and 57 in 1973. At this time, the population on Torishima has nearly tripled compared to the 1973 count, and, in addition, it has been confirmed that at least 25 other pairs are nesting on the Senkaku Islands.

The ideal environment for the nesting of the Torishima Island species is the volcanic-ash-covered hills, where uniformly large-tufted grass *(Miscanthus sinensis)* grows, not only holding together the soil but offering a minimum of shelter and support for the nests. By artificially inducing the spreading of this grass, the recovery of the primary population of the short-tailed albatross has been notably accelerated in the last few years. The island of Torishima is currently uninhabited by humans because in 1965 the meteorological station there was completely evacuated due to the threat of a giant volcanic eruption that, in reality, never occurred. However, two disastrous eruptions have already

With its white body, wings and tail almost entirely black, light-colored beak with a black point, and a sort of black mask around the eyes, the Abbott's booby does not have the chromatic characteristics of other tropical boobies and instead strongly resembles the Peruvian booby *(Sula variegata)*.

Its geographical distribution, however, is completely different from that of the Peruvian booby, given that it is found in the Indian Ocean and its nesting area is limited to Christmas Island. Here, the Abbott's booby nests in crowded colonies in the rain forests of the internal highlands, at altitudes of around 150 m (500 ft). Unfortunately, this nesting area coincides almost exactly with a zone rich in phosphate deposits, and mining operations have been under way on a large scale since 1965. These operations have, in effect, caused the destruction of vast areas of forest, leaving behind a lunarlike landscape. The boobies, for their part, are not capable of changing their habitats, and rather than move to new sites prefer to place their nests on the margins of the old ones, where there is still some degree of sparse vegetation. Regeneration of the rain forest seems very difficult, and the result may well be the complete loss of the one habitat remaining for this species' reproduction.

The Abbott's booby has a

However, more recent and accurate census taken found this figure to be an underestimate—in reality, there still exist at least 1300 nesting couples and around 5000 young boobies below the reproductive age.

The conservation problem regarding the Abbott's booby has become a classic testing ground for the coexistence of large economic interests and nature and conservation interests. Putting an end to the phosphate mining operations would assure the species a future, but that solution is not feasible because economic interests at stake are too great compared to the public opinion in favor of the boobies. Thus, a more difficult route is being attempted, that of exploiting the phosphate to the maximum extent while at the same time keeping harm to the boobies to a minimum. To this end, since 1975 a specialist has been placed on the island, charged with following the mining operations and furnishing instructions and advice for minimizing the disturbance these activities cause the birds. In practice, it is a question of paying the lowest possible price for this species' preservation by undertaking deforestation

The Abbott's booby nests in the rain forests of the internal highlands of Christmas Island.

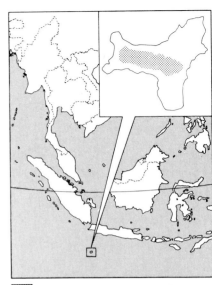

Distribution area of the
Abbott's booby

operations outside of the reproductive season and saving at least some stretches of forest where it turns out that phosphate concentrations in the land are not very high.

In this regard, an interesting initiative of Radio Australia is to be noted, which, taking a concrete interest in the fate of the Abbott's booby, has proposed and set up a reserve area of 400 hectares (899 acres) not very rich in phosphates. Before this simple plan was undertaken, the indiscriminate destruction of the forest between 1967 and 1974 caused the loss of no fewer than 120 nesting couples.

It should be noted, moreover, that until 1930 the Abbott's booby also nested on Ascension

The presence of valuable phosphates in the nesting area of the Abbott's booby presents complex problems for the protection of the species.

Island and that it was precisely mining operations for guano that caused its complete disappearance. The last wild specimen was observed on Ascension in 1936.

PHILIPPINES EAGLE

(Pithecophaga jefferyi)

Ex	R	E	T	V	I	K

This eagle, 97 to 98 cm (38 in) in length with a large crest of lanceolate feathers, is the ecological equivalent of the South American harpy. Although it does not have that bird's strength, it has a powerful beak, which is narrow, deep, and strongly curved; very strong legs partly covered by feathers; enormous feet and talons; and short but wide wings that, together with its long, square tail, constitute a perfect adaptation to forest life. Its flight, with flapping wings and short ascents, is unique to this predator, whose shape resembles that of a gigantic goshawk. The feathers of its head and long occipital crest are white and cinnamon colored, with shafts speckled dark brown. The feathers of the upper parts of its body and its wings are sepia colored, with a white and cinnamon border. Its long tail is dark brown with thin, almost indistinguishable stripes that are lighter and have a white tip. Its beak and cere are bluish black, its eyes azure, and its legs yellow. Its total appearance is unmistakable, in part because of the crest of feathers on its head. Unlike the harpy's crest, which stands erect, like a crown, the crest of the Philippines eagle falls backward

The erect head feathers on the specimen of Philippines eagle below reveal its state of excitement. On the following pages, the flight of the Philippines eagle through the forest.

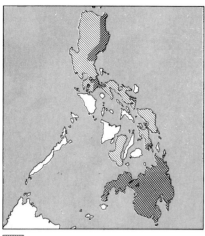

Distribution area of the Philippines eagle
Former area

like a long, thick mass of hair.

This eagle, proposed as a national symbol for the Philippines, is today in grave danger of extinction. Its distribution is already reduced to the humid and mountainous zones on only three of the archipelago's islands. Since 1934 there has been no word of it from Samar, where it once lived. On Luzon, it has been found in the Sierra Madre range in the provinces of Cagayan, Isabella, Nueva Vizcaya, and Quezon, though the most recent sightings were in 1963. Recently it has been rediscovered on Leyte, while on Mindanao it still has a vast range. During the last decade it has been observed in 15 of the 17 provinces of the island, but is by now missing from a large part of those flatland zones that are no longer covered by forest. Its population is in continuous decline.

In 1910, the population of this eagle in Mindanao was estimated to be 1200 individuals, this based on the extent of the existing forest (representing 65 percent of the entire island) and by assigning a territory of 100 square km (39 square mi) to each pair. Again with reference to Mindanao, where most of the population is presently found, between 1968 and 1970 the number of individuals was estimated to be only 30 to 60. However, after a careful reexamination of all the available data gathered until 1973, and using three different methods of calculation, the estimated number of specimens reached 309 to 580, with lower number in this range being probably the most reliable. Following these studies, it was also demonstrated that the territory occupied by each pair is, in reality, less than what was once supposed: 12 to 25 square km (5 to 10 square mi) In 1970, there were perhaps 8 to 10 individuals on Leyte, and although there is no precise information for Luzon, the population there is probably fairly small. In 1985 the total population was estimated to be below 300.

The preferred habitat of this species is primary evergreen forests—that is, those forests that have never been altered by man and that form a continuous, closed vault around 40 m (132 ft) from the ground and from which giant trees emerge here and there, sometimes as high as 60 m (198 ft)—from sea level to as high as 2000 m (6575 ft), though it also frequents certain secondary forests.

Because of its size and strength, it is one of the few species able to hunt monkeys (*Cynamologos*), and, in fact, *Pithecophaga* is Latin for "eater of monkeys." It also feeds on small dogs, pigs, and chickens, which it preys upon in villages near the forest. Individual pairs can become specialized in their diet and raise their young almost exclusively on one kind of prey, the selection being determined by which is the most accessible or vulnerable with respect to the nest's location.

Like other, similar eagles, the Philippines eagle builds its nest in the tallest trees, on the sides of mountains, or sometimes in the foliage of the gigantic kapok trees of the rain forest, trees very similar to those preferred by the harpy of tropical America. One can often see food leftovers hanging from the sides of their nests, and there is an accumulation of three or four thick layers of bones at the base of trees with nests, evidently a result of many years of continuous habitation. Even though only one eaglet is usually raised in each nest, two eggs may occasionally be laid, and one pair is known to have raised two young.

The species is completely protected: hunting, killing, or capturing it is illegal, and its capture to supply zoological gardens is strictly controlled. In spite of this, the eagle is shot down to supply trophies to private individuals and illegally trapped for zoos or private collections; each year from 5 to 10 percent of the population is removed in these ways. However, this reduction is far less than that caused by the destruction of its habitat. It is enough to consider that 65 percent of Mindanao was covered by forest vegetation in 1910, but that figure was down to 30 percent in 1973, and each year around 650 square km (250 square mi) of this island's forest is razed. For this reason, a selective felling of trees and reforestation using original tree species has been recommended to prevent the total destruction of the habitat suited to this eagle.

Several other useful suggestions have been put forward: the creation of reserves extending at least 200 square km (77 square mi) within which neither agriculture nor the cutting of trees would allowed, as for example in mountainous areas; the prohibition of the cutting of trees in national parks, reinforcing the laws the regulate the capture and holding in captivity of this species, and, finally, educating the local population on the importance of preserving the species, certainly one of the most necessary long-term measures to make this program of preservation effective.

Since many specimens are held in various zoos, both in the Philippines and in the rest of the world, exchanges among zoos

The Philippines eagle prefers primary evergreen forests, from sea level up to 2000 m (6575 ft); the species, in grave danger of extinction, has been proposed as the national symbol of the Philippines.

The Chinese egret has thus become an authentic local attraction, after having been mistreated and even ignored for many years. In fact, it reached a threshold very near to extinction for the simple reason that for a long time it was confused with another Asian species, the sacred egret *(Egretta sacra)*, which is still rather common. Only recently have zoologists realized that in reality it was a distinct, separate species and that it needed to be protected by special programs and arrangements.

Chinese egret on the Japanese island of Tsushima along the path of nesting sites in southern Korea. This migratory species nests in temperate zones and winters in tropical or subtropical regions.

ASCENSION FRIGATEBIRD

(Fregata aquila)

Ex	R	E	T	V	I	K

This species, characterized by very dark plumage (the males are completely black, the females have a dark gray chest), at one time nested in large numbers on Ascension Island in the Atlantic Ocean. For over a century, however, it has been completely extermi-

nated from that island by cats that arrived with European colonists, and it now nests only on the small island of Boatswainbird (5.5 hectares or 13.6 acres), which is separated from Ascension by a narrow (230 m or 756 ft wide) but turbulent channel that predators have never able to cross. Here, the frigatebirds nest on a totally bare slope that is part of a large mass of rocks with a covering of basalt.

At the end of the 1950s, the number of breeding pairs was put at 4000 to 5000, which would be a sufficient number were the nesting sites not so localized. Because the species continues to frequent the coasts of Ascension, it is believed that it could return there to nest if the cats were completely removed from the island. Until now, though, the control programs that have been undertaken have served only to temporarily reduce the number of felines. One hopes that access to Boatswainbird remains rigorously limited, seeing that it would be practically impossible to avoid the species' extinction if the predators managed to arrive there.

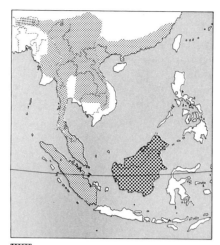

MILKY STORK

(Mycteria cinerea)

Ex	R	E	T	V	I	K

The milky stork of Southeast Asia is the only close relative of the painted stork, or Indian wood stork *(Mycteria leucocephala)*, a bird that is widespread in tropical Asia and that nests in spectacular colonies from India to southern China. The milky stork shares the characteristic shape of the painted stork, with a long, stout beak that is slightly curved near the end. Like the painted stork, the milky stork has bare skin of an intense red color on its face, but the milky stork does not have the painted stork's splendid colors. From that point of view, it more resembles the Asian open-billed stork *(Anastomus*

Former area
Current area

oscitans), being mostly gray and white with only the tips of the wings black.

Both stork species of the genus *Mycteria* nest in dense colonies in tall trees, often on islands with humid zones. They can sometimes be seen flying in spectacular V-formations or in long rows. These magnificent birds can often be seen in flight over urban areas, and, indeed, one colony of the more common *Mycteria leucocephala* is actually located in the middle of a city, in the zoological garden's park in Delhi.

The generic name *ibis* is sometimes used for these two stork species, with the painted, or Indian wood, stork referred to as *Ibis leucocephalus* and the milky stork as *Ibis cinereus*. However, aside from considerations of the rules of zoological nomenclature, the use of the term *Mycteria* is preferable because the word *ibis* creates confusion between these birds and the Ciconiiformes of the Treskiornitidae family, many which are called ibis in English, Italian, and other languages (but not in Latin).

The milky stork is probably still widely diffused in its vast habitat, which includes the southern part of the Malay

At right, a few milky storks. The species is believed to be well represented in Southeast Asia.

Among the many species in danger of extinction in various areas of our planet, some, for certain reasons, provoke more concern than others. This occurs when the species in danger of disappearing does not have close relatives that might in some way serve as evidence of at least some of the characteristics of the species that risks being lost. It also happens, of course, when the species in danger has clearly reached the final chapter of its odyssey—when the species is reduced to a population of a few dozen individuals or, in some cases, even fewer.

When these two reasons occur together—when a species with many unique characteristics finds itself in immediate risk of disappearing forever—the case arouses understandable concern and special attention, and the species is placed among the "priorities" for action by the associations that are concerned with the protection and conservation of nature and animals.

Among the species already irremediably extinct are many that, unfortunately, had more or less unique characteristics. The great auk *(Pinguinus impennis)* was the world's largest Alcidae, as well as being the only species that, like the Antarctic penguin, had lost the ability to fly. The dodo of Mauritius Island *(Raphus cucullatus)* was the only representative of the genus *Raphus* and one of the last two remaining of the Raphidae family, and since the other, the solitary of Rodriguez Island *(Pezophaps solitarius)*, has also become extinct, the disappearance of these birds has signalled the end of a very interesting group of uncertain parentage. Likewise, the parakeet of the Carolinas was the only species of parrot still existing in the Nearctic region as

peninsula, southern Cambodia, Southern Vietnam, and the islands of Sumatra and Java. Its numbers have decline a bit throughout its distribution area, however, because of the ongoing destruction of both the humid zones and the tree-filled sites suited for nesting. Another cause of its decline is the continued military operations that have tormented Indochina for many decades, damaging the nesting sites of this species in a way that cannot be ignored.

Little is known of the present status of this species. It probably still nests in at least one site in the state of Perak, in Malaysia, and also in Pulau Dua, along the northwest coast of Java.

Exact information concerning its distribution in Sumatra is missing, though its presence on the island seems certain, at least in the southern part. In any event, the milky stork is currently protected by laws both on Java and Sumatra, and the hunting of

The decline in numbers of this species is due to destruction of humid zones and the tree-filled sites suitable for nesting.

this bird is permitted, by necessity, only to those aborigines who live permanently in the forests.

well as the only representative of the genus *Conuropsis* existing on earth.

Clearly, the disappearance of a species with such special characteristics cannot be compared, with regard to the gravity of the loss, to the disappearance of one particular geographic form or subspecies of an animal that is otherwise widespread or even to the disappearance of a species that belongs to a genus of which there are other species with similar aspects or habits.

The case of the Japanese crested ibis must therefore be considered as one of the most grave and pressing dramas confronting a species today. Although this bird belongs to a family that still includes many, somewhat widespread members, it is the only one included under the genus *Nipponia*, and it is has a very unique appearance that is, in a certain way, intermediate between that of the sacred ibis, with its white plumage and completely bald head and black skin, and the bald ibis, with its brown plumage and almost entirely bare head with red skin and a tuft of feathers on the nape. The plumage of the Japanese crested ibis is entirely white, and the head is partially bare and red with a dense tuft of white plumes. This bird has been reduced to the lowest levels of a tragic downward spiral toward extinction, with only a dozen or so specimens still existing in Japan and an imprecise number, but certainly a very small one, in China. The fame and importance of its country of origin—developed countries with the necessary resources for conservation plans—have, up to now, helped focus attention on the bird from financially endowed groups and agencies, and this attention has prevented its extinction. There is no doubt, however, that the situation has not yet been resolved, and to survive, the Japanese ibis will have to have much luck and will have to find living conditions of a quality that thus far no one has been able to offer it.

Until the last century, the Japanese ibis was rather widespread in Asia, not only in Japan but also in China and the easternmost regions of the former Soviet Union. We also know that the continental population migrated toward the south during the harsh seasons, traveling as far as Hainan, Taiwan, Korea, and the Ryukyu Islands. The major areas of distribution of the Japanese population were the southern part of Hokkaido Island and the regions of Honshu and Kyushu.

The decline on the Asian continent began during the last century and was probably the result of hunting and the destruction of the habitat used for nesting. The last sighting of a nesting area in the former Soviet Union was in 1917, and the last known colony in Chinese territory was destroyed in 1952 when the woods in which the birds nested were completely cut down.

The sad tale of the species' decline in Japan begins in 1868, when, following a restoration of the empire, a series of political disorders took place that led to the concession of various reforms by the emperor. Among the old laws that were abolished was that concerning hunting, and the immediate result was a legalized persecution of many species of birds, among them this ibis. At that time, this bird was still common and widespread, but by 1890 every trace of the bird was lost, and for a period of around 40 years no further sightings of this bird were made, and the species was believed to have become extinct. In 1930, a very small population of this ibis was found nesting on the peninsula of Noto, and still others were found on Sado Island.

The Japanese ibis was accorded full legal protection in 1934, and also in the year the number of specimens still existing was put at around 50. During World War II, a good part of the forests

The number of Japanese crested ibises in Japan, China, and the former Soviet Union is believed to be fewer than 50. The ibises shown here flying over snow (below) may be the last survivors of the species in the Japanese archipelago.

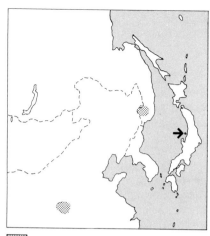

Distribution area of the
Japanese crested ibis
Breeding areas

that made up the habitat used by the species during the reproductive period was destroyed, and in 1960, when a new census was finally taken, it was found that barely 12 specimens remained. From that moment, the species has fluctuated around a total

In Japan, the last examples of the Japanese crested ibis live in a protected nature reserve on Sado Island.

number of 10, causing the fear, after each particularly severe winter, of a complete extinction (such winters are usually the cause of high mortality) and the hope, every time a brood is successfully raised, of survival.

The typical reproductive habitat of the Japanese ibis is pine woods, with the birds building their nests in the trees, while humid zones are the habitat the bird needs for feeding, but many such zones have been successfully converted to rice fields. The excessive use of pesticides in rice fields has also had a role in species' decline inexorably lowering the reproductive success rate of the few surviving couples. Fortunately, the rice fields on Sado Island that are now used as a feeding zone by the remaining ibises are relatively free of pesticides, thanks to a wise protection policy put into effect by the Japanese government. The Japanese government has also acquired the forests of Sado Island that the ibises still use for reproduction and, with the cooperation of the local World Wild-life Federation (WWF), plans to set up a series of facilities for the breeding of the species in captivity on the same island.

Because of various difficulties, this final project has not yet been launched, and the survival of the species is still hanging by a thread. In fact, even if one takes into account the fact that the number of specimens outside Japan is not well known and might be slightly higher than we think, the total number in Japan, China, and the former Soviet Union cannot be more than 20 to 50, and it seems very unlikely that this situation can be turned around in the near future.

WHITE-WINGED WOOD DUCK

(*Cairina scutulata*)

Ex	R	E	T	V	I	K

Above, a female white-winged wood duck with a brood

The so-called mute ducks (*Cairina moschata*), today raised all over the world domestically, are originally from South America and belong to the genus *Cairina* that includes at least one other species, *C. scutulata*. Unlike the wood duck, this species, originally from tropical Asia, has remained in a strictly wild state and is today threatened by extinction.

The white-winged wood duck is a rather stocky animal with large wings, short legs, and a head with irregular white dots. The female is smaller than the male, with a shorter neck and smaller head. The colors of the female's plumage are more opaque, and the white of its head is less evident. The best single feature for distinguishing between the two sexes is the color of the iris, which is dark brown in females but lighter and reddish or yellowish in the males.

Despite its large size and massive body structure, the white-winged wood duck has a surprisingly light and energetic flight. It flies among tree branches with great agility and often spends most of the day roosting in very tall trees (a strange habit for a duck).

The species' preferred habitat is humid, tropical forests cut by slow rivers and dotted with ponds protected by thick brush. It sometimes pushes itself as high as 1500 m (1644 ft) above sea level. During the rainy season it also frequents open swamps, but it deserts these during the rest of the year, because it does not like to expose itself to direct sunlight.

An omnivorous species, it feeds on seeds, aquatic plants, worms, frogs, mollusks, and small fish, its diet varying according to the season.

Its habitat has been greatly reduced during recent years because tropical forests have in large part been cut back in massive deforestation operations. For this reason, the species is today considered very rare. The naturalist Delacour holds that this is a question of a very shy animal that prefers to remain within thick vegetation and is thus difficult to observe. It is therefore possible that the population of this species is greater than believed. Once present in all of Southeast Asia, it has been drastically reduced, and today there are some 15 pairs living on six reserves in southeast Assam and two small groups of specimens in Bangladesh. It has disappeared totally from Thailand and Malaysia, but is still present in Sumatra, where it is even considered to be common in certain regions.

The ties formed between the members of mating pairs are weak: the male remains at times near the female and offspring, but its role in raising the young is not clear. The nests are built inside the natural hollows of older trees. During the reproductive season, this duck has a call, even during flight, that is very particular: it emits a sound, similar to that of a trumpet or a prolonged cry, that might best be defined as ghost-like. For this reason, local populations have nicknamed it the phantom duck.

The primary reasons for the worrisome numerical decline of this species are indiscriminate hunting of both young and adults, massive deforestation of the primary forest, and water pollution. The number of pairs now living in the wild is around 200. In 1971, the Wildfowl Trust began a program for the protection of this duck that provided for raising and reproducing it in captivity. Within 6 years, a good hundred specimens were raised in England. In 1978, several pairs were transferred to other collections, and the number of individuals now on breeding farms is fairly high, even in North America.

Unfortunately, when bred in captivity, this duck is subject to a form of aviary tuberculosis that is very worrisome. This occurs above all in England, where the climate is particularly humid. Direct sunlight can greatly reduce the incidence of this disease, but the species takes little advantage of the sun and prefers to remain in shaded areas.

International conventions presently in force should protect

the white-winged wood duck, which is protected in particular by law in Indonesia. A nature reserve of 2.59 square km (1 square mi) has been set up in Assam for its protection. The setting up of other reserves and the implementing of programs of breeding and reintroduction are the strategies now possible for the protection of this and many other species in grave danger of extinction.

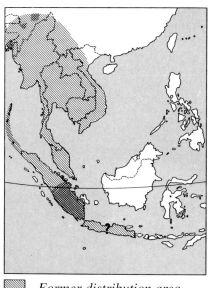

Below, a couple of white-winged ducks amid the vegetation of their habitat.

▨	Former distribution area
▨	Current distribution area
?	Possible areas of distribution

BULWER'S WATTLED PHEASANT

(Lophura bulweri)

Ex	R	E	T	V	I	K

The genus *Lophura* includes 10 species of typical pheasants found only in tropical Asia. Specialists consider them to be somewhat near the true wild roosters (genus *Gallus*), with whom they share a type of plumage that often tends toward silvery shades, a rather long tail (with central feathers that tend to form an arc), and a partly bare head with wattles and a crest (which, however, is in this case composed of feathers).

The genus *Lophura* includes some very common and widespread species, such as the silver pheasant *(Lophura nycthemera)* from Indochina, which breeds easily in captivity and is thus found in many zoos and private collections throughout the world. The genus also includes four species in danger of extinction: Edward's pheasant, the imperial pheasant, Swinhoe's pheasant, and Bulwer's wattled pheasant.

Bulwer's wattled pheasant is one of the few nonsedentary species in the family of Phasianidae. In fact, this pheasant regularly undertakes seasonal movements up and down the hills between 300 and 800 m (986 and 2629 ft) in search of fruit to eat. This species lives exclusively on the island of Borneo, and its distribution area is limited to a few places in the regions of Sabah, Sarawak, and Kalimantan. As early as 1910 its range was being broken up, and the species was rare in some areas and completely gone from others. The primary forest that makes up its natural

Left, above and below, the heads of two males of Bulwer's wattled pheasant. The blue wattles at the sides of the head become larger with age, as shown by the specimen on the right, which is older.

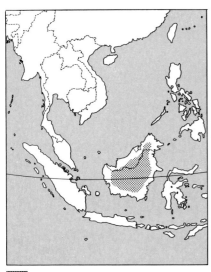

Mount Gunung Bondang, where the Bulwer's wattled pheasant is still to be found.

This species was imported to the West for the first time in 1976, and until now private collections have held specimens only sporadically. The first instance of breeding in captivity took place on a private breeding farm in Mexico in 1974. In 1992, at least 35 Bulwer's wattled pheasant were enriching American, European, and Mexican collections.

habitat has been the subject of intense exploitation for many years; not even the national parks and reserves has been spared. The only exceptions are the national park of Mount Kinabalu (712 square km or 275 square mi), that of Puala Gaya (1276 hectares or 3153 acres), and several other small tracts of primary forest. A third of the forest of Kutai (3060 square km or 1181 square mi), had already been cut down by 1977, and the exploitation of oil deposits in the park's interior is foreseen. It has been predicted that if the government does nothing to block the intensive exploitation of timber, all of the primary forest in Kalimantan will be converted to agricultural land by 1995.

The Bulwer's wattled pheasant is only one of many species that live in and depend on that particular habitat. Thus, this drastic alternation of a habitat will lead to extremely serious consequences for many of the island's fauna. The only protection that this Phasianidae enjoys is the prohibition of capturing live specimens for international commerce. Another proposal, put forward to safeguard at least part of the forest patrimony, is the creation of a reserve in central Kalimantan, in the zone called Bakit Raya, which would include

BROWN-EARED PHEASANT

(Crossoptilon mantchuricum)

Ex	R	E	T	V	I	K

The brown-eared pheasant lives in the mountains of southern Chacar, in Mongolia, and in the Chinese provinces of Hopei and Shansi. It was believed to have disappeared from Hopei, but then a specimen from that province was sold in Peking in 1933. Even so, there has been no certain news on the status of this species since 1949.

The habitat of the brown-eared pheasant is mountain forests, shrub lands, and prairies. The only measures taken to protect this species have been the control of its commerce in those countries belonging to CITES, since the brown eared pheasant was included in Appendix I in 1973. It is to be hoped that certain zones are soon set up where the release of specimens bred in captivity can be undertaken; indeed, reproduction on farms was already taking place during the last century, and in 1864 a male and two females were sent to the Paris Zoo, where they successfully reproduced in 1866. Then, the London Zoo received two other males, and all the pheasants of this species now held in captivity (1000 specimens) originated with these pairs. Another two pairs were sent to the Pheasant Trust from China in 1976.

The brown-eared pheasant lives in the mountains of southern Chacar, in Mongolia, and in the Chinese provinces of Hopei and Shansi. It was believed to have disappeared from Hopei, but then a specimen from that province was sold in Peking in

There is little recent information on the situation of the brown-eared pheasant, a bird of mountain forests, shrub lands, and prairies.

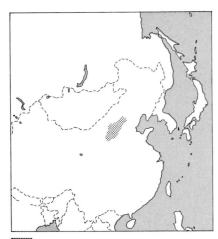

Distribution area of the brown-eared pheasant

1933. Even so, there has been no certain news on the status of this species since 1949.

The habitat of the brown-eared pheasant is mountain forests, shrub lands, and prairies. The only measures taken to protect this species have been the control of its commerce in those countries belonging to CITES, since the brown eared pheasant was included in Appendix I in 1973. It is to be hoped that certain zones are soon set up where the release of specimens bred in captivity can be undertaken; indeed, reproduction on farms was already taking place during the last century, and in 1864 a male and two females were sent to the Paris Zoo, where they successfully reproduced in 1866. Then, the London Zoo received two other males, and all the pheasants of this species now held in captivity (1000 specimens) originated with these pairs. Another two pairs were sent to the Pheasant Trust from China in 1976.

EDWARD'S PHEASANT

(Lophura edwardsi)

Ex	R	E	T	V	I	K

The Edward's pheasant is a rare species found only in a very narrow area in the provinces of Quang Tri, Hue, and Fai Fo in Vietnam, just to the south of the old border between North and South Vietnam. The male of this species is very similar to that of the imperial pheasant and is almost entirely metallic purple-blue with a short crest of feathers on its head that, unlike that of the imperial pheasant, is white (the imperial pheasant's crest is the same color as the rest of its plumage). The female's plumage is a chestnut brown with surrounding feathers bordered in a lighter brown, giving the bird a scalelike appearance. The similarity of the Edward's pheasant to the imperial pheasant (females of the two species are practically indistinguishable in the wild) may reflect a true affinity: both species are believed to be recent derivations of the Salvador pheasant *(Lophura inornata)*, which lives on Sumatra and, because of the slight specialization of its plumage, it considered to be an ancestral form of the *Lophura*.

Because of the narrowness and particular location of its distribution area, the Edward's pheasant has always been considered a species difficult to observe in the wild, and today little is known of its behavior and biology in the wild. The species lives mainly in the intricate thickets of secondary regrowth, where it is very difficult to make sightings, and even when

The habitat of the Edward's pheasant is the humid forests on the hills of three provinces of Vietnam. In the vegetation that has regrown after the use of defoliants during the Vietnam War, the Edward's pheasant seems too have reached promising numbers.

a sighting is made, it is usually very short.

The species is believed to have suffered serious damage during the Vietnam War, and the devastation wrought by the massive use of defoliants has without doubt upset its habitat and, at least locally, caused the death of many wild animals. Even before the beginning of the war, the Edward's pheasant was rare and localized in the most humid forests, those with dense underbrush of liana covering the irregular terrain of Vietnam's calcareous hills. American soldiers stationed in Vietnam are reported to have sighted the species fairly regularly and to have killed it often to obtain an unscheduled meal. Despite this, a report from 1975 related that the Edward's pheasant was recovering in the new vegetation that has regrown after the use of defoliants and that it had even attained a numerical consistency unheard of in the past.

Above, an example of a female Edward's pheasant. The species is present in various zoos and breeding farms throughout the world.

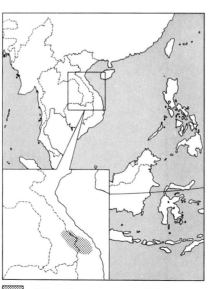

▨ *Distribution area of the Edward's pheasant.*

Even so, the species was still listed in Appendix I of the CITES and is regularly kept and bred in captivity for cautionary reasons. The World Pheasant Association has prepared a book *(Studbook)* on the kinships of all the specimens raised and has recommended to the Vietnamese government that it protect the species in the wilds as a "national treasure."

A pair of Edward's pheasants was exported to the West for the first time in 1924. One year later, it was breeding with success. Since, then, other specimens from this species captured in the wild have been sent to various zoos and breeding farms in Europe and North America, where they have regularly bred. In 1976, surveys taken of individuals of this species held in captivity in Europe and North America revealed 635 specimens. It is interesting to note that despite the close kinship between the Edward's and imperial pheasants, the two species attain sexual maturity at different times; upon reaching 2 years of age for the imperial and 1 year for the Edward's. Nevertheless, the two species have been crossbred, and absolutely inadvisable practice when dealing with animal in danger of extinction, which, in the last analysis, should be reintroduced into the wild with all their original characteristics and where, moreover, the habitat requires well-defined characteristics from the genetic point of view in order to allow for adaptation and survival.

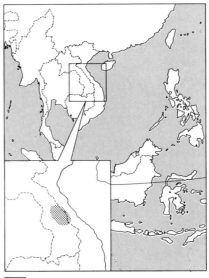

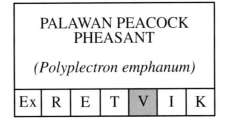 *Distribution area of the imperial pheasant*

IMPERIAL PHEASANT

(Lophura imperialis)

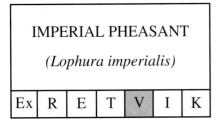

Ex	R	E	T	V	I	K

The presence of the imperial pheasant in the Vietnamese provinces of Dong Hoi and Quang Tri and in several areas in Laos has been confirmed. This species' situation is not known with certainty because gaining access to those areas was, for political reasons, virtually impossible in the past and because the vegetation in those areas is difficult to penetrate. This is believed to be a case of a small population that is diminishing.

The major cause of its decline is the serious alteration of its habitat caused by the defoliants used by US military forces during the Vietnam War. Many animals were poisoned by these herbicides, and although the effects on the imperial pheasant are not documented, the poisoning and destruction of its habitat have made this species vulnerable.

The typical habitat of the imperial pheasant is forests and mountain underbrush. To save this species, the Pheasant Trust of Norwich has instituted a program of reproduction in captivity with the object of eventual reintroduction of this species into the wild.

PALAWAN PEACOCK PHEASANT

(Polyplectron emphanum)

Ex	R	E	T	V	I	K

Peacock pheasants are relatively small Galliformes characterized by plumage marked by a large number of colored ocelli ("eyes") that, during the nuptial display, are exhibited in a splendid "wheel" composed not only of the quill feathers of the tail but also of the covering feathers of the wings, which are spread out with the head lowered.

The Palawan peacock pheasant, also called the Napoleon peacock pheasant, is the most beautiful species of the genus *Polyplectron*. The feathers of the upper parts are of a metallic blue changing to green, with golden reflections, while the wings are similar in color to the back, though lighter. The head, adorned with a long occipital tuft, is green with azure reflections. The back and tail are dark, banded and spotted in a ruddy color. Each of the quill feathers is adorned toward the apex with two ocelli of a metallic green color changing to gold, while the

primary covering feathers have only one ocellus. The eyes are banded by a white eyebrow; also white is the posterior extremity of the cheeks as far as the covering feathers of the ear. Unlike the males, the female has a uniformly brown color and ocelli only on the quill feathers. The head of both sexes has an upright tuft. According to the zoologist Ghigi, two subspecies can be found in the same area that differ from one another only by the absence or presence of the white eyebrows. As Ghigi says, "It is a question of a second genotype that cannot manage to separate itself from the first because of the lack of geographical isolation." In 1959, Ghigi imported from the Philippines a pair of each of the two subspecies, and he crossed the male with white eyebrows with the female from the other subspecies. All of the hybrid offspring—5 chicks born in 1961 and 10 in 1962 (of which seven survived)—had the white eyebrows.

The Palawan peacock pheasant lives only on the Philippines island of Palawan. Its habitat is the primary forests of the coastal plains and part of the more arid wooded areas at the foot of the hilly zones. Although there is no firm supporting information, it seems that this species is adapting itself to the secondary forest that has been regenerated after the primary was cut down. The Palawan peacock pheasant is supposedly totally protected by law, but in reality it is hunted intensively or captured alive for export (a 1974 survey revealed that around 100 were captured each year). The decline of this species can also be attributed to the shrinking of its habitat, and thus the fate of the Palawan peacock is tied to its ability to adapt to a more open environment. If it turns out that the species is not completely dependent on the forest, its future should be relatively secure. Other-wise, the only way to assure its survival would be the institution of new protected areas.

The Palawan peacock pheasant was exported for the first time in 1929, to California in the

United States, and it was there that the first specimens were born in captivity; the first pair to reach Europe came from California in 1931. In 1992, the number of peacock pheasants in captivity amounted to more than 350.

A pair of Palawan peacock pheasants is shown. The male, with its showy plumage, is behind; in front is the female.

Distribution area of the Palawan peacock pheasant
Area where the species is probably extinct.

<table>
<tr><td colspan="8">SWINHOE'S PHEASANT
(Lophura swinhoii)</td></tr>
<tr><td>Ex</td><td>R</td><td>E</td><td>T</td><td>V</td><td>I</td><td>K</td></tr>
</table>

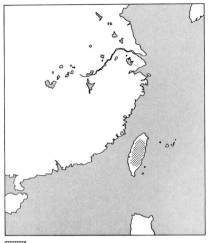

Distribution area of the Swinhoe's pheasant

According to certain authors, the Swinhoe's pheasant can be considered, together with the Edward's pheasant and the imperial pheasant, as one of the group of so-called blue-rumped pheasants. This color characteristic distinguishes the males of the three species and reflects a relationship even closer than that which one might normally expect among species belonging to the same genus. Perhaps in this case the concept of a semispecies or superspecies could be applied, thus making clear a common origin that is more recent than that which can be hypothesized for the other members of the genus *Lophura*, as for example the silver pheasant. The blue-rumped pheasants display undoubted affinity with the silver pheasant, as can be seen in the fecundity of their relative hybrids, but this relationship is less direct than that which the three blue-rumped species present among themselves. With respect to the other Lophura, they show many differences, not just in appearance but also in behavior.

Once spread uniformly over the entire island of Taiwan, the Swinhoe's pheasant is today found mostly in the mountains of that island and in the hilly area in the vicinity of Taipei. It prefers primary broadleaf deciduous forests covering mountainous ridges no steeper than 40°, but also inhabits mature secondary forests. Unfortunately, the zones below 2000 m (6575 ft) are undergoing profound changes during these years because of changes in how the land is exploited. Today, in place of the woods and forests, there are agricultural and silvicultural holdings, hydroelectric plants and mines, tourist sites and road networks associated with these activities. The increasing transformation of these mountain areas and the greater ease of access for humans create increasingly worrisome disturbances for the wild fauna. As far as the Swinhoe's pheasant is concerned, given the very limited range in which it still survives, one can assume that it has become somewhat rare, even if one cannot rule out the possibility that it has in some way adapted to the new environmental conditions.

This species was protected by Japanese law from 1919 until the end of World War II. In 1967, the Taiwan government declared it completely protected, and a 1974 law was passed that prohibited the exportation of wild fauna. A further form of protection, begun in 1972, is the prohibition of hunting a certain number of species, among them the Swinhoe's pheasant. Also in 1974, the Taiwan government set up a reserve of 3680 hectares (9093 acres) at high altitude, with the primary intention of protecting this pheasant and the Mikado pheasant (*Syrmaticus mikado*), which is also threatened with extinction.

In 1967, five females and six males born in captivity at the Pheasant Trust of Norwich, Great Britain, were released in the experimental forest of Che Chi, which belongs to Taiwan's National University, near Hitsou, in the inland mountains. Another six pairs were released in the same zone the following year. These birds were subsequently sighted many times within the area of their release; thus, one can hope that this species may be saved from extinction by repopulating suitable zones.

The Swinhoe's pheasant is listed in Appendix I of the CITES; as a result, the commerce of these birds is under strict control in the nations adhering to the convention. In addition, it has been proposed that the number of national parks and nature reserves be increased and the those already existing be enlarged. In particular, five zones particularly well-suited to this Phasianidae have been pointed out

to the Taiwan Forestry Bureau and the National Council of Sciences.

Right, a female Swinhoe's pheasant; its plumage is more sober and camouflaged than that of the male. Below, the blue plumage of the male Swinhoe's pheasant joins it to the Edward's and imperial pheasants in the group of so-called blue-rumped pheasants.

ELLIOT'S PHEASANT

(Syrmaticus ellioti)

Ex	R	E	T	V	I	K

The immense continent of Asia can be defined in many ways, even if looked at from only the naturalistic point of view. It would not be totally incorrect—although it might seem somewhat limiting—to define it as the continent of pheasants. Out of around 190 species of the family, 108 are found in Asia; and of the true pheasants—excluding the francolins, partridges, and guinea fowl—the Asiatic species represent practically all of them. Unfortunately, this situation is also reflected on the IUCN *Red List*: of 26 species of Phasianidae in danger of extinction, 18 are of Asian origin. They are also among the most beautiful birds in the world.

Some authors call the Elliot's pheasant the chessboard pheasant because its plumage is composed of large white, black, copper, and steel-blue squares. This species is characterized by the absence of a tuft on its head, by its much developed and upright carrades, and, in the males, by its very long tail made up of 16 quill feathers, the middle ones appearing much longer and more pointed than the others. The females have a much paler coloration than the males and also have a shorter tail. Young males and females can also be distinguished since the middle quill feathers of the young males are adorned with light-color crosswise stripes alternating with chestnut-colored streaks. In young females, the tail is uniformly grayish.

Egg laying begins toward the middle of the year, and an average of 10 eggs is laid, rarely more. At the end of the incubation period, which lasts 25 days, the chicks emerge from the small, rose-colored eggs, covered by down adorned with chocolate color stripes.

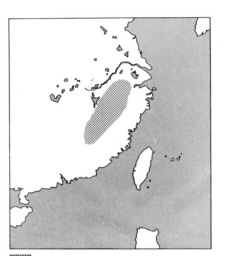

Above and right, Elliot's pheasant. Because of the coloration of its plumage, the species is also called the chessboard pheasant.

▨ *Distribution area of the Elliot's pheasant*

The distribution area of the Elliot's pheasant includes the Chinese provinces of Kiangsi, Anhwei, Chekiang, Fukien, and northern Kwangtung. The species' preferred habitat is dense forest, interrupted by steep ravines, not at a high altitude. The forests of eastern China have been subject to vast destruction, and what remains is ever more accessible to hunters and poachers, who have undoubtedly also played an important role in the decline of this species. The numerical consistency of the species in the wild is not known, but as early as 1910 it was described as "not common" and localized. In 1951, it was considered to be rather diffuse, though not abundant. It is believed that since then its numbers have fallen further, in part because its habitat has in large part been destroyed.

The Elliot's pheasant is listed in Appendix I of the CITES and is also legally protected. There has been talk of initiating a program of reintroduction, making use of animals raised in captivity, on condition that protected natural habitats suited to this species are put at its disposal.

The Elliot's pheasant was discovered in 1872 by the naturalist Swinhoe in Chekiang and southern Anhwei. The next year it was found again by Father David at Kuatun in northern Fukien. In 1874, Father David exported to Europe the first specimen of this species, sending one from Kuatun to Paris, where it lived for many years at the menagerie of the local museum.

The first couple to reproduce

in captivity did so in Europe in 1880, and as far as we know, the principal zoos of the world now possess more than 120 specimens, while private collections hold as many as 1500. Thus, this species can count on a numerous population, kept in cages, which may serve as a useful security system if its preservation in the wild is not handled with the necessary attention.

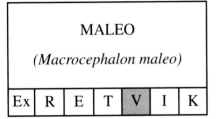

MALEO

(Macrocephalon maleo)

| Ex | R | E | T | V | I | K |

The only species of the genus *Macrocephalon*, the maleo is the most vividly colored representative of the entire family of Megapodiidae. Its tails and wings are black, its lower parts rosy, and its head is ornamented by a helmet of feathers. Like all Megapodiidae, it is a terrestrial species, rather shy, and rarely takes flight. Its distribution area is composed of the Indonesian island of Sulawesi; moreover, it was probably introduced during the last century to the Kepulauan and Sangihe islands, which are between Sulawesi and the Philippines.

The birds usually live in the primary forest of the plains, but for nesting they choose open areas, glades or beaches, where the ground is sandy and easily heated by the sun or underground hot-water springs. Like all Megapodiidae, the maleo has evolved a unique incubation method for its eggs, one that permits a notable saving of energy. The female lays each egg in a deep hole (around 1 m or 1 yd) in the sand, then covers the hole and allows the incubation to take place

Usually a forest bird, the maleo chooses open, sandy areas or beaches for nesting.

Distribution area of the maleo

through solar or volcanic heating. This expedient is only the first stage of a much more complex system adopted by other species of the Megapodiidae family, such as the *Megapodius freycinet* (which inhabits the dense tropical forests of Polynesia and the Philippines), which lays its eggs in a mound of earth mixed with vegetable matter that ferments, creating the necessary heat. Such "incubating piles" can reach 5 m (16 ft) in height and have a diameter of around 11 m (36 ft). The eggs are laid in tunnels, 1 m (3 ft) in length, that runs through the pile. Many pairs cooperate in the construction of these large nests, in which the temperature oscillates between 30° and 38°C (86° and 100°F).

This method of incubation doesn't protect the eggs from predators. In the case of the maleo, the eggs are easily dug out by animals such as wild pigs and by man. For this reason, many maleo populations have suffered drastic reductions in their populations. For example, the colonies on Panua, Imandix, and Honayono and, in particular, a colony on Batu-putih, disappeared in 1919, which was 6 years after an indigenous community settled in that area. Thirteen colonies of maleo were counted in 1979 in the island's northern part, each composed of from 30 to 1250 adult individuals. On the assumption that each female lays 30 eggs a year, an estimate was made of a

presence of more than 3000 individuals on the northern part of Sulawesi and of 5000 to 10,000 on the entire island.

The species as a whole has been protected by Indonesian law since 1970. In reality, the young and adults have been protected since the 1930s, but their eggs were still subject to uncontrolled exploitation. In fact, the government transferred the management of the beaches used by the maleo for nesting to the communities of local villages for the harvesting of eggs.

Several reserves have been built on Pulau Garat (at the extreme north of the Kepulauan island chain), on Tangkoko-Batuangus, Panua, Manembo Nembo, Bone, Dumoga, and Lore Kalamanta. Moreover, another area in the southeastern part of Sulawesi, where several maleo populations were found, has been protected. A 3-year ecological-conservationist study was begun in 1977 and sought, among other things, to analyze the possibility of reconstructing several colonies by transferring and reburying eggs in the areas occupied in the past by the maleo. But, above all, it is important that a rigorous control is begun to avoid the excessive exploitation of this species.

On the IUCN *Red List* is another species belonging to the family of Megapodiidae, *Megapodius laperouse*, which is listed as rare. Two subspecies of this are known: *Megapodius laperouse laperouse* and *Megapodius laperouse senex*. The first is found on almost all the Mariana Islands in the western Pacific. Its presence has been noted on the islands of Asuncion, Agrihan, Pagan, Alamagan, and Tinian. Its habitat is made up of woods of low vegetation, whether along beaches or on hills. Data are not available on the population's density, but it is known that goats, wild pigs, and reptiles constitute a serious threat. This megapode is completely protected by law, but the problem remains of the Mariana natives, who prefer to eat this bird's eggs rather than its meat.

The subspecies *M. l. senex* lives on the Palau Islands in the southwestern Pacific. It can be found in the internal forests of the largest

islands as well as in rocky areas on small, calcareous islands. The population must surely have been more numerous in the past. Estimates taken in 1945 on four islands indicated the presence of barely 45 to 70 specimens. Though more recent information does not exist, the fact that this megapode is spread out on different islands offers promise for the survival of this subspecies.

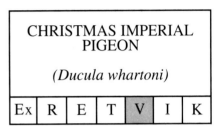

CHRISTMAS IMPERIAL PIGEON

(Ducula whartoni)

Ex	R	E	T	V	I	K

Endemic to Christmas Island, south of Java, the Ducula whartoni is part of a group of imperial pigeons that inhabits the small islands of Oceania, where they appear to be isolated from the other species of this genus. The fall in the species' population, already underway at the beginning of this century, is today furthered by phosphate mining that, besides damaging the habitat, permits the arrival of hunters in places that until now hosted this species.

Several scientific expeditions between 1965 and 1974 estimated a remaining population of fewer than 100 pairs, a population perhaps slightly growing. It is not rare today to see this bird in the rain forests of the highlands in the island's interior, particularly in the tops of high trees, where it feeds on large fruits, usually during the last hours of daylight.

A proposal was recently made to create a nature reserve to protect the habitat of another bird threatened with extinction, the *Sula abbotti*, and such a program would benefit the *Ducula whartoni*.

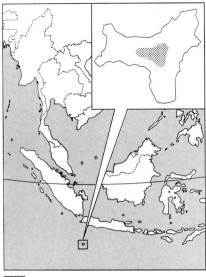

Distribution of the Christmas imperial pigeon

TRISTRAM'S WOODPECKER

(Dryocopus javensis richardsi)

Ex	R	E	T	V	I	K

This woodpecker is in serious danger; although a protected species—it is considered a true "natural treasure"—it is perilously close to extinction and can be found in only a few locations in central Korea. According to various authors, from 5 to 14 subspecies of this woodpecker have been described, with distribution areas covering a large part of southern Asia; except for this one, none of these birds is running any risk of extinction.

Like most of the so-called black-and-white woodpeckers, the Tristram's woodpecker is

Below, a male Tristram's woodpecker perched near its nest, dug in the trunk of a tree. Although protected, the species is very near extinction.

large. In fact, it can reach 46 cm (18 in) in length. In this subspecies endemic to Korea, adults have the crown of their heads and several spots on their cheeks a vivid red, their upper parts, throat, and the upper part of the chest black, and the rest of their lower parts and back white. Adolescents are recognizable by their white-speckled throat.

At one time, this woodpecker was found on the small island of Tsushima, located in an arm of the sea between Korea and Japan. The population was never very large here, because there has never been a sufficient forest covering on the island to sustain more than several pairs. Its extinction in this area is, nevertheless, almost surely to be attributed to collectors, who, heedless of its small population, have always captured or shot down all specimens on sight. In fact, the last woodpecker of which information was had on the island was taken in 1920. Now, the only remaining hope is that some individuals from Korea may return to the island.

This woodpecker is still found in several provinces in the northern part of South Korea, in particular at Mount Solak and Kwangneung, for which, however, there are no precise numerical estimates, and in the southern part of North Korea, where, after a very fast decline, and increase in the population was registered from 20 individuals to around 30 to 40 a few years later in 1969.

Even in the past, however, this woodpecker has never been particularly abundant, since a large area is necessary to maintain a single pair. Even so, there is no doubt that its population has notably suffered from the extensive destruction of the forests that has been going on for a long period of time in this part of the world. In particular, the most serious environmental disaster occurred during the Korean War in the 1950s, which resulted in the extinction of many species and the serious danger in which many others now find themselves.

The destruction of trees, carried out in a systematic way,

→ *Distribution area of the Tristram's woodpecker*

Former distribution area

fered severe damage, as a result not only of the war but also of the deforestation carried out to make room for farms and roads.

Part of its habitat is protected in South Korea in that it falls within the borders of nature parks or reserves. In North Korea, several zones where the subspecies nests have been declared areas to protect. There is, for example, a Tristam's woodpecker reserve at Rinsan. Even so, in neither of the two countries have other protective measures been considered. A project is underway for raising the species in captivity.

using herbicides to deprive the enemy of cover, has spelled out doom for these woodpeckers, making it impossible for them to find trunks within which to build nests and thus regularly reproduce. The hopes for the survival of the Tristam's woodpecker now rest with those birds that have survived in zones that remained outside the major combat.

This woodpecker's preferred habitats are mature forests of pine, fir, oak, and camphor; certain preference has been noted in North Korea for building nests in trunks of chestnut trees. Unfortunately, most of the region's mature forests have suf-

Below, a museum example of the Tristram's woodpecker, captured on the island of Tsushima.

LANYU SCOPS OWL							
(Otus elegans botelensis)							
Ex	R	E	T	V	I	K

This subspecies is considered in danger of extinction because it is found only on the island of Lanyu, 29 km (18 mi) from the coast of Taiwan, where its favored habitat, the forest vegetation, has been nearly entirely destroyed by the intense development for tourism and replaced in large part by introduced vegetation.

Toward the end of the 1960s, this scops owl was considered still rather common on the slopes of the island's central chain, but a study carried out in 1973 in this area's remaining forest gave negative results. Specimens could be heard calling to one another in only one of the wooded canyons. In 1985, however, the presence of this bird on the island was confirmed. Although it seems that some pairs are able to survive in what remains of the forest vegetation, it is difficult to imagine that they will be able to continue to do so for much longer if some kind of intervention is not undertaken.

This subspecies has been legally protected since 1974.

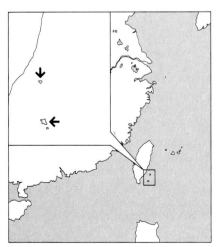

⬅ *The Lanyu scops owl and its distribution area:*
⬇ *Former area*

OKINAWA WOODPECKER							
(Sapheopipo noguchii)							
Ex	R	E	T	V	I	K

The sparse and sporadic distribution of this woodpecker, which is the only representative of the genus *Sapheopipo*, is restricted to a small hilly area on the island of Okinawa, one of the Ryukyu group south of Japan. The devastating effects of World War II and continual deforestation have contributed to reducing the extension of its preferred habitat, the wooded areas; but there are also other factors that still contribute to harming it, such as uncontrolled fires and the substitution of the original arboreal vegetation with new, exotic plants.

As early as 1920, this species was considered very rare, and in 1941 its population was estimated to be below 100 individuals. In 1973, it was estimated that there were no more than 60 and not fewer than 20 remaining pairs. Different recent counts have confirmed the existence of around 100 specimens.

Its preferred habitat is the humid forests of broadleaf evergreens, in particular the primary and undisturbed forests, although it sometimes frequents secondary forests in the search for food. In this region, the primary forests have by now been reduced to small islands surrounded by secondary forests, plantations, or deforested terrain. Moreover, these are constantly threatened be deforestation, carried out to harvest wood or plant other cultivations, and—in the summer—by fires.

A large part of the forests have been used by the US Marines, who maintain a base on the island, even though its jurisdiction has recently been placed again under Japan; in these areas, however, the Marine Corps has always discouraged the cutting down of trees by the local population. After the violent demonstration by the ecology movement against military exercises within the range of the species, the Japanese government in 1964 also acquired private land in order to increase the protected zones. Unfortunately, though, only 36 percent of the territory is protected; it will thus be necessary to enlarge these areas and block further deforestation or, when this is inevitable, to see that it is performed in a way that leaves the largest trees, thus allowing these woodpeckers nesting sites.

In 1973, this species was declared a "natural treasure." A nature reserve of 7 hectares (17 acres) was instituted on Mount Yohana, where two or perhaps three pair are found. Other zones on mounts Ibu and Nashime are controlled by the forest department of Ryukyu, but since they are fenced in, the protection is in effect only nominal.

The forests of Okinawa in which this species is found:
 *Natural forests*

Prefecture forests

Private forests

Right, nesting Okinawa woodpecker. Already considered rare 70 years ago, this species is today thought to number about 100 specimens.

The Mukojima Bonin honeyeater (the only species belonging to the genus Apalopteron) is a Meliphagidae found on the Japanese islands of Bonin, which are locally known as Ogasawara. Two subspecies have been identified, the nominal (Apalopteron familiare familiare), which originally populated the islands of Muko, Nakodo, and Chichi in the northern part of the Bonin archipelago, and that of Hahashima (A. f. hahashima), which is limited to six islands in the Hahashima archipelago, which, in turn, is found in the southern area of the Bonins.

The members of both sexes of this species are very similar in appearance: they have yellow heads, the chest and stomach have triangular black markings that are also found in a T-shape on the face around the eyes; the

Birds that nest in evergreen forests, the Mukojima Bonin honeyeater has suffered a disastrous decline due to the destruction of its forests habitat caused by changes made for the tourist trade. Below are two specimens drinking.

134

back is olive, the beak and claws are black, the iris is brown, and a thin ring of naked white skin surrounds the eyes.

The differences between the two subspecies are quite small: all that can really be noted is a more intense yellow coloration and a smaller size in the nominal subspecies.

One of the most interesting characteristics of this bird is the structure of its tongue, which is typical of the Meliphagidae, with its tip divided into thin filaments, similar to those of a brush, very useful for extracting nectar from flowers. Aside from this trait, this bird is not really similar to other birds in its family. The behavior of the Mukojima Bonin honeyeater has not yet been studied in all its details, but it is

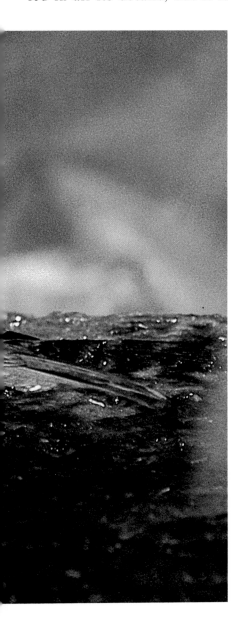

known that these birds are notably social: they move in groups of 10 to 20 specimens in inland forests and in those along the coast, making their cup-shaped nests among the lower branches of trees (1 to 3 m or 3 to 10 ft from the ground). They usually lay from two to four eggs in the month of May. The duration of the brooding is about 2 weeks, and the young birds become capable of providing for themselves at the age of about 3 months.

These birds nest in evergreen (camellia, wild orange, etc.) forests, but outside the reproductive period they also visit green areas near inhabited areas. They obtain their food using a variety of interesting systems: moving along the ground capturing insects and eating fallen fruit, capturing insects in flight or extracting them from the bark of trees, hanging onto the trees like woodpeckers. They also go out onto the branches of trees to eat the pulp of mature fruit, along with the nectar of flowers. They

have a strong and melodious song. In a study published in 1979, the nominal subspecies was for the first time judged to be in danger of extinction. It seems, however, that at that time it was already extinct, given that 11 years earlier, during the autumn of 1968 (on the occasion of the restitution to Japan of the archipelago, which at the end of World War II had remained under American occupation) an exploration had been conducted without any sightings.

The Hahashima subspecies, however, still survives and is found on the islands of Muko and Haha; however, it, too, would seem to have become completely extinct on the island of Mei in 1940 and also on the island of Imoto, where it was originally present only in small numbers. The reasons for this disastrous decline are first of all the destruction of the evergreen forests of *Peucaena*, *Pandanus*, and *Hibiscus*, which have been in large part cut down to make space for a growing tourist

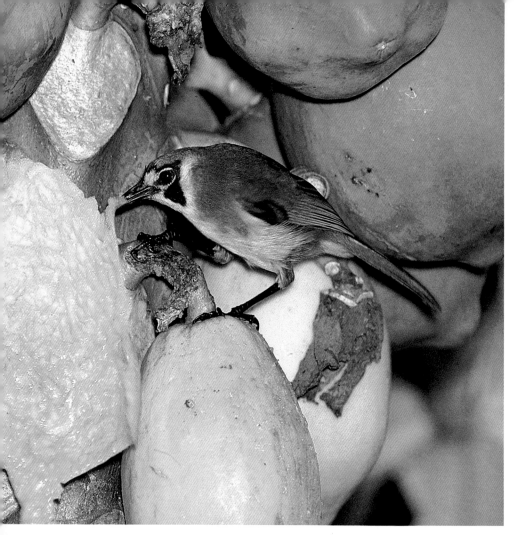

ROTHSCHILD'S STARLING

(*Leucopsar rothschildi*)

This rare and threatened species, considered in danger of extinction, was discovered only in 1910; its distribution is confined to an area of around 200 square km (77 square mi) in the northwestern part of Bali (Indonesia) and to a small island off its southern coast, Nusa Penida.

The Rothchild's starling, which attains a total length of 28 to 33 cm (11 to 13 in), has a yellow bill and legs, an almost completely white plumage, apart from the black on the wing tips and tail, and has an area of bare skin around the eyes colored a deep blue. The two sexes are very similar, but only the male has long plumes on the nape that extend down the back. Because of its very beautiful appearance, this species is actively sought after by breeders, who are willing to pay high prices to have them in their collections.

In 1925, hundreds of this species were counted in their areas of greatest concentration, but in 1976, during the course of a study carried out in nine localities, equal to around one third of their range, 127 individuals were counted. It was thus deduced that the total population must have been around 500 individuals, and was in any event certainly less than 1000.

Several persons have suggested that the excessive number of birds captured to satisfy the demand of collectors should be considered the main cause of the decline of this species. However, it has been shown that it was cutting down of forests, carried out to make room for human settlements, that caused the confinement of this stupendous starling in a narrow area of its former range. Thus it is easy to imagine how the number has more likely decreased due to the continuous

Above, a Mukojima Bonin honeyeater eating papaya; the bird is prevalently an insectivore.

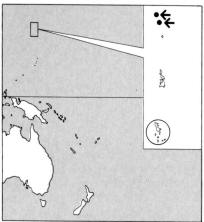

Distribution area of the Mukojima Bonin honeyeater:
→ *Apalopteron familiare familiare*
○ *Apalopteron familiare hahashima*

mortal competition with it.

The Mukojima Bonin honeyeater was declared by Japanese law a "natural treasure" in 1969 and was made a bird due special protection measures in 1972. Up to now, there has been no news of concrete efforts undertaken to protect what little remains of this bird's habitat.

industry; second of all is the introduction of the mejiro, a small *Passeriforme zosteropide* that occupies an ecological niche very similar to that of this honeyeater and that thus enters into a

136

diminution of its preferred habitat.

In fact, in 1925 the western part of Bali, more than one third of the island, was, with the exception of a few towns, completely covered by primary forests. In 1976 there were still substantial areas of forest that were once occupied by this starling, but the changes in the forest cover that in any event accompa-ny human settlements had already encouraged and favored the spread of a species closely related and fairly similar in appearance, the *Sternus melanopterus*. This bird, which in 1926 was not even present within the range of the Rothschild's starling, during the 1970s came to be around three times more numerous than this species, even in those zones where the most relevant populations are present, that is, in areas of uncontaminated forest, in

The capturing of specimens by collectors and the cutting down of the forests have confined the rare species of Rothschild's starling, discovered in 1910, to a small zone on the island of Bali.

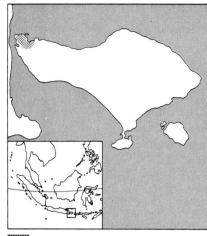

wheat fields, and in coconut plantations: in effect, those zones on Bali least changed by man.

Since 1957 the species has been protected in Indonesia by an ordinance for the protection of nature, and since 1971 its capture has been expressly forbidden, as well as its being shot down and its exportation. Despite this, this splendid animal continues to be illegally captured and exported to satisfy the demand of collectors. Part of its range is included within the Bali-Barat Natural Reserve (20,000 hectares or 44,942 acres), and the population of that area, around 4500 persons, is supposed to be moved elsewhere, according to what has been established by the government. The breeding of this species has been underway for a while. The first breeding in captivity occurred in 1931. In 1964, 10 percent of birds belonging to collections had been born in captivity, and in 1973 this figure had risen to 27 percent, and in 1974 to 36 percent. There has already been news of successive generations in several zoos. On the basis of this information it can be stated that the species is in all likelihood attaining levels of reproduction in captivity that are capable of completely satisfying the demand, thus rendering totally useless the capture of wild birds.

JAPANESE CRANE

(Grus japanensis)

Ex	R	E	T	V	I	K

The Japanese crane, or tancho, as it is called in Japan, is among the most elegant and legendary of cranes. Once found throughout Japan, this species is now in serious danger of extinction and survives only in small colonies in China, in the eastern part of Japan's Hokkaido Island, and in certain zones in Siberia.

This bird is of notable size (around 1.5 m or 5 ft tall) and has particularly long legs (the tarsus attains 30 cm or 12 in in length) that are thin and black. Its plumage is mostly snowy white with long black remiges and two large black areas extending along both sides of the head, neck, and throat. The top of its head is bare, as is the case with many Gruidae, and is a vivid red; rough, black skin covers the forehead and cheeks. The white primary feathers and tail hide long, black secondary feathers.

To the Japanese, the tancho is a symbol of joy and longevity: according to popular legend, this bird lives a thousand years. It has been immortalized by centuries of Japanese and Chinese painters and is a recurring figure in art, literature, and mythology. Its image is found almost everywhere and is even the symbol of the Japanese airline.

This species breeds in the easternmost regions of Asia, in the Khanka Lake basin in the former Soviet Union, and along the tributaries of the Ussuri and Amur rivers. Several small colonies also breed in northern and central Manchuria and China. In Siberia, this species nests in the extensive plains and swamps covered with reeds of *Calamagrostis*. In Japan,

Japanese cranes courting. On the following pages, spreading the wings is part of the courting ritual.

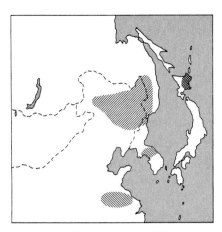

Distribution area of the Japanese crane:

Nesting zone

Wintering zone

it makes use of habitats characterized by open waters, reeds of *Phragmites*, and small clearings in alder and sedge woods. From January to March, when the reproductive period has begun, the Japanese crane exhibits itself in spectacular courtship rituals: the pairs, walking in snow-covered fields, begin their dance by repeatedly bowing, arching their wings and curving them along their backs, raising their necks and heads. The movements gradually become faster, and the two partners move about more quickly, flapping their wings, rising and holding themselves in flight as far as 3 m (10 ft) from the ground, and giving off sharp calls very similar to hoarse blasts from a trumpet.

Until a few years ago, this crane's courtship ritual was considered a manifestation of happiness, an occasion during which the two partners harmoniously collaborate for the recreation of their species. This interpretation is no longer widely accepted; more generally accepted is the view that the courtship and mating activities show signs of the conflict of interest between individuals of the two sexes, individuals that achieve a difficult alliance according to which each tries to transmit the largest quantity of its genes to the next generation. The role of this courtship is certainly complex. It seems, for example, that in those species in which both care for the young, the female uses these com-

plex rituals to evaluate her companion's strength and ability at raising young.

The courtship rituals are repeated several times, then the pairs return to nest in the same place as the preceding year. Each pair has a vast breeding territory, from 2 to 7 square km (1 to 3 square mi), in which the birds build a somewhat tall nest of grass and intertwined branches. The eggs, usually two, are brown with brown and gray spots and are brooded in turn by the two parents for about one month. The young hatch at the end of April or beginning of May and are covered by a brownish red down. In the beginning, they are very clumsy, but within 9 to 10 weeks they abandon the nest and, when they have finished shedding, they are ready to follow the adults in their migrations.

The population on Hokkaido is sedentary during the winter; in fact, these cranes distance themselves only a few dozen kilometers (about 25 mi) from the swampy zones where they were born in the spring. They are the only cranes that winter on snow. They perch in trees during the night, glide over the snow-covered fields in the early morning, and spend the day poking their beaks from time to time into the snow, searching for grains of wheat that have remained in the ground after the preceding year's harvest.

The Japanese crane has been hurt by relentless hunting by collectors and by the destruction of its habitat. Between 1942 and 1945, in the northern Chinese province of Hopeh, small flocks of some 50 specimens were seen in spring as well as during their autumn migrations. In 1964, 30 to 35 pairs were spotted on the beaches of Khanha Lake, and the entire Soviet population numbered from 200 to 300 specimens. In South Korea, the Japanese crane appeared to be totally absent between 1953 and 1962, though 41 individuals were present on the Han River in the winter of 1974-1975; 50 were counted in the vicinity of Panmunjom, and 20 at Inchon; in the winter of 1977-1978, the population of Chorwon numbered 100. Today, a small colony of Japanese cranes breeds regularly in the

vicinity of the Pyongyang Zoo in North Korea.

This Gruidae was very common in Japan before 1890. Its population has since undergone a strong decline. During the 1920s the colony breeding in the swamps of Kushiro, in western Hokkaido, reached its historical minimum— no more than 20 were counted. In Japan in 1933 the first programs for the species' protection began with the creation of reserve areas and the distribution of food during the winter. The food project was made successful with the valuable help of farmers and school children and was supported by private contributions and subsidies from the local government. The success of these measures has been clear; in 1960, the number of species amounted to 175 and increased in subsequent years so much so that in the winter of 1976-1977 the number of wintering specimens in Hokkaido was 220, of which 40 were offspring.

The Japanese people are very sensitive to the fate of this crane, probably because the species has a fundamental role in many traditional legends. It was perhaps for this reason that this bird was made a natural treasure in 1935 and a "special natural treasure" in 1952. Today, the species is protected through international conventions among Japan, the United States, and the former Soviet Union. Moreover, a reserve of more than 5000 hectares (12,000 acres) has been created in the Kushiro swamps (the Japanese Crane Protected Area), but, unfortunately, only three pairs, or 5 percent of the population on Hokkaido, breed in this area. The remaining 95 percent breed in zones that are subject to intense exploitation by man or that have been earmarked for agricultural or industrial uses, or that will be crossed by new roads in the near future.

The Japanese government recently proposed widening this reserve so as to include within it two large areas and bring the number of protected breeding pairs from 3 to 52. In 1958, another protective area for the species was established, also in the Kushiro swamps (Kushiro Natural Crane Park). Today, this area is frequented by about 20 specimens.

In North Korea, the Japanese crane is protected by law, and its wintering area in Hwanghai province was proclaimed a national treasure in 1968. The same was done for the estuary of the Han River in South Korea. The hunting of this species is prohibited in the Soviet Union.

One of the major causes of death of the Japanese crane in Japan (accounting for 12 percent of the species' total per year) is the high density of electrical wires in the air. Cranes frequently fly into these and are electrocuted. For this reason, it has been proposed that electrical lines in the vicinity of the grazing areas of Shimmosettsuri be placed underground or, at least, moved away.

At the moment, the fate of this Gruidae is not closely tied to the presence of specimens raised in captivity; in fact, the number of specimens in the wild appears sufficient to guarantee the survival of the species. On the other hand, it is comforting to know that the species is raised in numerous zoos and private collections throughout the world, for the most part in China, Korea, and Japan. Moreover, the Japanese crane reproduces in captivity without great difficulty.

Another crane still exists in the same habitats, but in western China, and although it is not as rare as the Japanese crane, it finds itself in serious danger; this is the black-necked crane *(Grus nigricollis)*. This is a bird of remarkable size (around 120 cm or 47 in tall), with gray plumage over its entire body except for the black primary and secondary remiges, the bare, red top of the head, and the dark gray tail bordered in a lighter gray. Of course, this species' most obvious characteristic is the black neck that gives the species its name. This species breeds in Tibet, from Ladak to Koukou Nor, east of Tsinghai, on large islands of inland lakes or

Japanese crane in flight. Gravely damaged by hunting until the end of the last century, it was considered extinct until its rediscovery in 1923.

along banks rich in vegetation, as high as 4000 m (13,148 ft). The nest is a somewhat deep and cup-shaped, located in moss and made of aquatic plants and grass gathered nearby. The eggs, one or two in number, are laid in May: they are olive brown with reddish spots.

Only in recent years has this species begun to decline sharply; in fact, it is cited in the updated IUCN *Red List* of birds in danger of extinction. This decline makes the situation truly worrisome. It must be observed that public concern for wild animals is still insufficient, so much so that action to protect a species is taken only when that species is in grave danger, and not before, when respect for these animals would help them survive.

It is also true that the most serious danger for the future of the Japanese and black-necked cranes is exploitation and reclamation of humid zones—that is, of the zones that are the primary habitat of these fascinating species. Thus, the most important effort that man should undertake is the protection and preservation of these areas.